Building Technology 2

LEEDS COLLEGE OF BUILDING LIBRARY
NORTH STREET
LEEDS LS2 7QT
TEL: 0532 430765

Building Technology 2

Jack Bowyer
Dipl. Arch. (Leeds) Architect

Illustrations prepared by
Peter Bowyer
Dipl. Arch. (Brighton)

THE BUTTERWORTH GROUP

UNITED KINGDOM
Butterworth & Co (Publishers) Ltd
London: 88 Kingsway, WC2B 6AB

AUSTRALIA
Butterworths Pty Ltd
Sydney: 586 Pacific Highway, NSW 2067
Also at Melbourne, Brisbane, Adelaide and Perth

CANADA
Butterworth & Co (Canada) Ltd
Toronto: 2265 Midland Avenue, Scarborough,
Ontario, M1P 4S1

NEW ZEALAND
Butterworths of New Zealand Ltd
Wellington: T & W Young Building,
77–85 Customhouse Quay 1, CPO Box 472

SOUTH AFRICA
Butterworth & Co (South Africa) (Pty) Ltd
Durban: 152–154 Gale Street

USA
Butterworth (Publishers) Inc
Boston: 19 Cummings Park, Woburn, Mass. 10801

First published 1978

© Jack Bowyer, 1978

All rights reserved. No part of this publication may be reproduced or transmitted in any form or by any means, including photocopying and recording, without the written permission of the copyright holder, application for which should be addressed to the publisher. Such written permission must also be obtained before any part of this publication is stored in a retrieval system of any nature.

This book is sold subject to the Standard Conditions of Sale of Net Books and may not be re-sold in the UK below the net price given by the Publishers in their current price list.

British Library Cataloguing in Publication Data

Bowyer, Jack
  Building technology.
  2.
  1. Building
  690   TH145   77-30407

ISBN 0–408–00299–9

Typeset by Butterworths Litho Preparation Department
Printed in England by The Whitefriars Press Ltd,
London and Tonbridge

Contents

INTRODUCTION

MODULE A FIRST FIXING JOINERY AND WINDOWS 1

1.00   First fixing joinery and its protection 1
     1.01 Wet applied finishes 1
     1.02 Timber grounds and arris beads, etc. 1
     1.03 Floors 3
     1.04 Timber for joinery 3
     1.05 Protection of timber from decay etc. 3
     1.06 Moisture content of timber for joinery 4
     1.07 Drying of timber 4
     1.08 Protection of joinery — general 4
     1.09 } Knotting and priming, etc. 4, 5
     1.10
     1.11 D.p.c.'s to external frames 5
     1.12 Protection from physical damage 5
2.00   Windows and window boards 6
     2.01 Machine preparation of joinery timbers 6
     2.02 Surface reduction tolerances 6
     2.03 Steel windows to BS 990 6
     2.04 Aluminium windows to BS 4873 7
     2.05 Requirements for windows 7
     2.06 } Building Regulations concerning
     2.07   windows 7
     2.08 Types of window and opening lights 7
     2.09 Performance requirements for windows 9
     2.10 EJMA stormproof window 9
     2.11 Vertical sliding sash windows 10
     2.12 Steel windows 10
     2.13 Aluminium windows 10
     2.14 Selection of materials for window boards 10
     2.15 Materials in use for window boards 10
3.00   Door frames and linings 11
     3.01 Construction and fixing of frames and linings — general 11
     3.02 Strength of frames and linings 12
     3.03 } Weather exclusion to frames and thresholds 12, 13
     3.04
     3.05
     3.06 Internal door linings 13
     3.07 Internal door frames for fire check doors 14
     3.08 Fixing door frames and linings 14
4.00   Staircases and landings 14
     4.01 } Building Regulations concerning stairs and landings 14
     4.02
     4.03 Planning and layout of domestic staircase 15
     4.04 Trimming stair well openings 15
     4.05 Basic staircase construction — timber 16
     4.06 Staircase strings, spandrels and carriage piece 16
     4.07 Balustrades and handrails 17
     4.08 Winders 18
     4.09 Half and quarter landings 18
     4.10 Geometrical staircases 18

MODULE B SERVICES AND DRAINAGE 19

5.00   Services — general introduction 19
6.00   Public utility services and intakes 19
     6.01 Water supplies 19
     6.02 Main water collection and distribution 19
     6.03 } Laying water mains 19, 20
     6.04
     6.05 Connections to supply mains and communication pipes, 20
     6.06 Water supply pipes into buildings 20
     6.07 Materials for pipes 20
     6.08 Stop and drain cocks 20
     6.09 Electricity supply 20
     6.10 Underground supply into buildings 21
     6.11 Overhead supplies into buildings 21
     6.12 Sealing chambers, meters and fuse boards 21
     6.13 Gas supply 21
     6.14 Sewage disposal by public sewer 21
     6.15 Principles of operation 21
     6.16 Restoration of roadway surfaces 22
7.00   Hot and cold water installations 22
     7.01 The Direct cold water supply system 22
     7.02 The Indirect cold water supply system 22
     7.03 Hot water supply — alternative heat sources 23
     7.04 } Types of system in use 23, 24
     7.05
     7.06 Materials 24
     7.07 Pipework in copper including sizing 24
     7.08 Control valves 24
     7.09 Fittings for pipe connections 25
     7.10 Cold water storage cisterns 25
     7.11 Ball valves 25

7.12 Overflows 26
7.13 Hot water cylinders 26
7.14 Boilers 26
7.15 Insulation 27
8.00 Installation of sanitary fittings 27
    8.01 Types of sanitary fitting 27
    8.02 W.C. 28
    8.03 Bath 28
    8.04 Lavatory basin 29
    8.05 Sink 29
9.00 Heating domestic structures 29
    9.01 Problems of poor combustion 29
    9.02 ⎫ Building Regulations concerning boilers,
    9.03 ⎭   hearths and flues 29, 30
    9.04 Construction of traditional open fire and hearth 30
    9.05 Precast and prefabricated units for open hearth and flue construction 32
    9.06 ⎫ Trimming and weathering structure to
    9.07 ⎭   allow passage of flues 33, 34
    9.08 Construction of traditional chimney cappings 34
    9.09 Flues for gas fired appliances 35
10.00 Domestic foul and stormwater drains 36
    10.01 Factors affecting layout of domestic drains 36
    10.02 Building Regulations concerning drainage 37
    10.03 Drain pipes 38
    10.04 Vitrified clayware pipes 38
    10.05 Cast iron pipes 39
    10.06 Pitch fibre pipes 39
    10.07 Unplasticised PVC pipes 40
    10.08 ⎫ Drain laying 40
    10.09 ⎭
    10.10 ⎫ Laying rigid pipes with flexible joints
    10.11 ⎭   41
    10.12 ⎫
    10.13 ⎬ Inspection chambers – design and con-
    10.14 ⎭   struction 41, 42
    10.15 Interceptor traps 42
    10.16 Ventilation of drains 43
    10.17 Soil and ventilating stacks – single stack system 43
    10.18 Sizes of pipes for small installations 43
    10.19 Design of soil pipes 43
    10.20 Stormwater drains – principles of collection and disposal 44
    10.21 Soakaways 44
    10.22 Testing drains 45

MODULE C FINISHES AND FINISHINGS 46

11.00 Finishes and finishings – definition 46
    11.01 Wet and dry finishes, etc. 46
    11.02 Internal finishes 46
12.00 Wall plastering, dry lining and external rendering 46
    12.01 Problems of background, suction and movement 46
    12.02 Plastering 46
    12.03 ⎫ Movement, cracking and mechanical
    12.04 ⎭   damage 47
    12.05 Batching and mixing of plaster 47
    12.06 ⎫ Types of plaster on solid backgrounds
    12.07 ⎭   47
    12.08 Lath and plaster on timber stud partitions 47
    12.09 Preparation of backgrounds to improve suction or adhesion 48
    12.10 Quality of finished plastering 48
    12.11 Plaster cored coves 48
    12.12 Dry lining with gypsum plasterboard 48
    12.13 Types of plasterboard 48
    12.14 External renderings – general 49
    12.15 Preparation of background and application of 2 coat rendering finish 49
    12.16 Tyrolean renderings 49
13.00 Ceramic wall tiling – general 50
    13.01 Typical backgrounds 50
    13.02 ⎫ Methods of tile fixing 50
    13.03 ⎭
    13.04 Grouting and cleaning wall tiles 50
14.00 Ceiling plastering and finishes 50
    14.01 ⎫
    14.02 ⎬ Plaster lath and wall board – sizes,
    14.03 ⎬ thickness and fixing, etc 50, 51
    14.04 ⎭
    14.05 Application of plaster to plaster board 51
    14.06 Patent textured applications 51
15.00 Floor screeding and finishes 51
    15.01 Preparation of concrete base 51
    15.02 Maximum areas for laying screeds 52
    15.03 Laying cement and sand screeds 52
    15.04 Curing cement and sand screeds 52
    15.05 Defects and their treatment 52
    15.06 Floor finishes – general requirements 52
    15.07 Floor finishes – laying techniques 52
    15.08 Floor finishes – dimensions and thicknesses 53
    15.09 Expansion joints 53
    15.10 Cure and protection of floor finishes 54
16.00 Applied decoration and painting 54
    16.01 Problems of substrates 54
    16.02 Preparation and application of paint 54
    16.03 Knotting resinous surfaces 55
    16.04 Priming – application to various substrate materials 55
    16.05 Filling nail holes, etc. 55

16.06 Primers, undercoats and finishing coats 55
16.07 Water bound paints and emulsions 56
17.00 Glass and glazing 56
    17.01 Sheet glass 56
    17.02 Translucent glasses 56
    17.03 Float glass 56
    17.04 Wired glass 56
    17.05 Double glazing 57
    17.06 } Glazing, putty and mastics 57
    17.07 }

MODULE D SECOND FIXING JOINERY AND DOORS 58

18.00 Second fixing joinery 58
    18.01 Programming with other trades 58
    18.02 General fixing principles 58
    18.03 Doors – hanging 59
    18.04 Hinges – butt and strap (tee) 59
    18.05 Hinges – other special patterns 60
    18.06 Selection, preparation and protection 61
19.00 Architraves and skirtings 61
    19.01 Standard stock designs 61
    19.02 } Jointing 61
    19.03 }
    19.04 Priming and sealing 61
20.00 Function and construction of basic doors 62
    20.01 Performance requirements of doors 62
    20.02 Timber quality and moisture control 62
    20.03 }
    20.04 } Development of the traditional board door 62, 63
    20.05 }
    20.06 Panelled and flush doors – general 64
    20.07 Panelled doors to BS 459 Pt. 1. 64
    20.08 Flush doors to BS 459 Pt. 2. 65
    20.09 Ironmongery 65
21.00 Sliding doors and gear – general types 66
    21.01 Side fixing patterns 66
    21.02 Soffit fixing patterns 67
    21.03 Cupboard doors 67
    21.04 Glass panel doors 68
22.00 Fire check doors and frames in timber – situations and general requirements 68
    22.01 Building Regulations concerning fire resisting doors and frames 68
    22.02 Door conforming to BS 459 Pt. 3 (Half and one hour fire resistance) 69
    22.03 } Wood frames for above 69, 70
    22.04 }
    22.05 Hinges and ironmongery 70

MODULE E SITE WORKS, ROADS AND PAVINGS 71

23.00 General site works and subsoil drainage 71
    23.01 Site works 71
    23.02 Subsoil drainage – reasons for provision 71
    23.03 Systems in use 71
    23.04 Design procedure 72
    23.05 Drainage pipes 72
    23.06 Diversion of ground water around a building 72
    23.07 Disposal of collected ground water 73
    23.08 Breaking in of drain trenches 73
24.00 Domestic and estate footways and pavings – reasons for provision 74
    24.01 Forms of construction 74
    24.02 Removal of stormwater or natural drainage 74
    24.03 Gullies 74
    24.04 Open joints 75
    24.05 Excavations 75
    24.06 Sub-base materials 75
    24.07 Edging materials 75
    24.08 Granite sets and cobbles 76
    24.09 Surfacing – alternative materials 76
25.00 Estate roads and drainage 76
    25.01 Alternative constructions 77
    25.02 DOE Regulations and gradients 77
    25.03 Excavation of roads and evaluation of subsoil 77
    25.04 Footway crossings 77
    25.05 Stabilisation of subsoil on sloping sites 77
    25.06 } Road gulley pots and gratings 78
    25.07 }
    25.08 Kerbs 78
    25.09 Carriageway sub-base and surfacing 79
26.00 Reinstatement and landscaping 79
    26.01 Clearance of building waste 79
    26.02 Reinstatement of top soil 79
    26.03 Edging to grassed areas 79
    26.04 Turves and turfing 80

Bibliography and references

MODULE A

BS 1186 Pt. 1 and 2: 1971 *Quality of timber and workmanship in joinery*
BS 4072: 1974 *Wood preservation by water borne compositions*
BS 1336: 1971 *Knotting*
BS 990 Pt. 2: 1972 *Steel windows*
BS 4873: 1972 *Aluminium alloy windows*

MODULE B

CP 310 *Water supply*
CP 304 *Sanitary pipework above ground*
CP 305 *Sanitary appliances*
BS 3505: 1968 *uPVC pipe for cold water services*
BS 1972: 1967 *Polythene pipe for cold water services*
BS 2871 Pt. 1: 1971 *Copper tubes for water services, etc.*
BS 1010: 1973 *Stop valves for water services*
BS 1212 Pt. 1 and 2: *Ball valves*
BS 417: 1973 *Galvanised m.s. cisterns, etc.*
BS 2777: 1974 *Asbestos cement cisterns*
BS 4213: 1967 *Polyolefin or olefin copolymer moulded cold water cisterns*
BS 699: 1972 *Copper cylinders for domestic purposes*
BS 1566 Pt. 1: 1972 *Copper indirect cylinders – single feed*
BS 1952: 1964 *Copper alloy gate valves*
BRS Digest 15 (2nd series) *Pipes and fittings for domestic water supply*
BRS Digest 80 (2nd series) *Soil and waste systems for housing*
BS 416: 1973 *Cast iron spigot and socket soil waste and ventilating pipes*
BS 4514: 1969 *uPVC soil and ventilating pipes*
BS 1184: 1961 *Copper and copper alloy traps*
BS 3943: 1965 *Plastics waste traps*
BS 1213: 1945 *Ceramic wash down w.c. pans*
BS 1125: 1973 *W.C. flushing cisterns*
BS 1188: 1974 *Ceramic wash basins and pedestals*
BS 1189: 1972 *Cast iron baths*
BS 1244 Pt. 1: 1972 *Metal sinks for domestic purposes*
BS 1454: 1969 *Consumer's electricity control units*
Institution of Electrical Engineers: *Regulations for the Electrical Equipment of Buildings*
BS 1181: 1971 *Clay flue linings and terminals*
BS 1289: 1945 *Precast concrete flue blocks for gas fires*
BS 715: 1970 *Sheet metal flue pipes for gas fired appliances*
CP 131.101: 1951 *Flues for solid fuel domestic appliances*
CP 337: 1963 *Flues for gas appliances up to 150 kW rating*
CP 302: 1972 *Small sewage treatment works*
CP 301: 1971 *Building drainage*
CP 308: 1974 *Drainage of roofs and paved areas*
BRE Digest 151: *Soakaways*
BS 65 and 540 *Clay drain and sewer pipes and fittings*
BS 539: 1971 *Dimensions of fittings for use with clay drain and sewer pipes*
BS 556 Pt. 2: 1972 *Concrete pipes and fittings*
BS 437: 1970 *Cast iron drain pipes*
BS 2760: 1973 *Pitch fibre drain pipes and fittings*
BS 4660: 1973 *uPVC drain pipes and fittings*
BS 1130: 1943 *Cast iron drain fittings*

MODULE C

CP 211: 1966 *Internal plastering*
BRE Digest 1960 *External rendered finishes*
CP 221: 1960 *External rendered finishes*
BS 1191 Pt. 1: 1973 *Gypsum building plasters*
　Pt. 2: 1973 *Premixed light weight plasters*
BS 12 Pt. 2: 1971 *Ordinary portland cement*
BS 890: 1972 *Building lines*
BS 1198: 1955 *Sands for internal plastering*
BS 1199: 1955 *Sands for external renderings*
BS 1230: 1970 *Gypsum plasterboard*
BS 1369: 1947 *Metal lathing for plastering*
BRS Digests 49 *Choosing specifications for plaster*
BRS Digests 104 *Floor screeds*
Gypsum Plasterboard Development Association: *Dry lining (1961)*
British Gypsum Ltd: *White Book (1972/3)*
CP 212 Pt. 1: 1963 *Wall tiling – Internal*
CP 212 Pt. 2: 1966 *Wall tiling – External*
BS 1281: 1966 *Glazed ceramic wall tiles and fittings*
British Ceramic Tile Council Technical Specifications for ceramic wall tiling
See text (15.08) for bibliography on floor finishes
CP 231: 1966 *Painting of buildings*
BS 2521: 1966 *Lead based priming paints for wood*
BS 2523: 1966 *Lead based priming paints for iron and steel*
BS 2525–7: 1969 *Undercoating and finishing paints (white lead based)*
BRS Digest 55, 56, 57 *Painting walls*
BRS Digest 106 *Painting woodwork*
BRS Digest 70 *Painting iron and steel in buildings*
CP 152: 1972 *Glazing and fixing of glass for buildings*
BS 952: 1964 *Classification of glass for glazing*
BS 544: 1969 *Linseed oil putty for timber frames*

MODULE D

BS 1331: 1954 *Builders' hardware for housing*
BS 455: 1957 *Locks and latches for doors*
BS 1227: 1945 *Hinges*
BS 584: 1967 *Wood trim (softwood)*
BS 459 Pt. 1: 1954 *Panelled doors*
BS 459 Pt. 2: 1962 *Flush doors*
BS 459 Pt. 3: 1951 *Fire check flush doors, wood and metal frames*
BS 459 Pt. 4: 1971 *Match boarded doors*

MODULE E (see also Module B)

BS 1194: 1969 *Concrete porous pipes for subsoil drainage*
BS 1196: 1971 *Clayware field drain pipes*
BS 340: 1963 *Precast concrete kerbs and edgings*
BS 594: 1973 *Rolled asphalt for roads and paved areas*
BS 1446: 1973 *Mastic asphalt for roads and footways*
BS 1447: 1973 *Mastic asphalt (limestone) for roads and footways*
BS 802: 1967 *Tarmacadam for roads (crushed rock aggregate)*
BS 1241: 1959 *Tarmacadam for roads (gravel aggregate)*
BS 1242: 1960 *Tarmacadam for footpaths, etc.*

BS 1621: 1961 *Bitumen macadam for roads and paths*
BS 368: 1971 *Precast concrete flagstones*
BS 435: 1931 *Granite and whinstone setts*
BS 539: 1971 *Fittings for use with clay drain or sewer pipes*

BS 556: Pt. 2: 1972 *Concrete manholes, inspection chambers and street gullies, etc.*
BS 497: 1967 *Cast iron manhole covers, gulley gratings and frames.*

Introduction

The introduction of the Technician Education Council Standard Units for Construction Technology has caused the whole basis of teaching in this subject to be drastically revised. This is due to a number of factors the principal being

(a) the time available for class contact and evaluation has been reduced to 60 hours per session. This is approximately half that previously available for ONC teaching in this subject;

(b) much of the old syllabus has been collated and included in separate subject units and therefore the wider scope for explanation and practical project work is now no longer controlled within the time available for the Construction Technology Unit.

In the preparation of *Building Technology 1 and 2* the extent to which the range of alternative solutions to construction problems should be carried has been carefully considered. It has been decided to omit much of what has, in the past, been considered as general practice and to concentrate on expanding and detailing good sound modern practice. For example, the construction of brick turned arches and accompanying centering has been omitted, together with first floor fireplace construction. In their place opportunity has been taken to explain the problems of condensation and its avoidance, to incorporate a number of modern timber jointing and strengthening techniques and to include modern methods of flue construction for stoves and boilers. Other divergencies from traditional techniques will be apparent to the informed reader.

The two volumes which, within the parameters outlined in this introduction, cover the first two years' syllabus of the student in Building and Construction Technology.
Building Technology 1 covers the introduction to the subject and the construction of the superstructures of small domestic and single storey steel framed buildings.
Building Technology 2 deals with the provision and installation of joinery, services and simple drainage, finishing trades and external roads and footpaths and preparation for landscaping.
It is considered that by preparing the books in this way the logical development of a building can best be described to the junior student.

Reference has been made to other TEC Units. Where essential to the proper understanding of a constructural element a brief explanation is included in the relevant module to integrate the two; technical aspects being omitted in favour of the specific subject unit. It is therefore essential that the following units be considered both complimentary and essential to the proper understanding of Building Technology.
1. *Construction Drawing 1 (TEC U75/049)*
2. *Site Surveying and Levelling 2 (TEC U75/056)*
   In respect of the preparation of site surveys and levels and detailed setting out by optical methods.
3. *Organisation and Procedures 1 (TEC U75/071)*
   Includes details of contracts and contractural documentation, British Standards and Codes of Practice, general information on the building professions and building legislation.
4. *Science and Materials 2 (TEC U75/042)*
   Includes detailed consideration of a wide variety of subjects including theoretical problems of sound and thermal insulation, structure and defects of wood, selection of aggregates, manufacture and testing of concrete defects and strengths of bricks and blocks, plasters and paint films and the theory of corrosion.

The construction of modern buildings is almost wholly controlled by the directions and 'deemed to satisfy' conditions contained within the Building Regulations and parallel legislation. As it is essential that the student should, from the beginning of his studies, understand and relate these requirements to his constructional studies care has been taken to relate the text directly to the relevant paragraphs in the Regulations e.g. (D3)* general requirements on foundations.

In addition, wherever possible previous references to detailed explanation are cross referenced in the text by noting the particular paragraph e.g. (7.03) Hot water supply.

Changes in technician education have also taken place in Scotland with the introduction of Scottish Technical Education Council (SCOTEC) certificate courses for building technicians. These courses have not adopted the modular approach of TEC but they do broadly include the same subject matter. Scottish students will find these two volumes cover practically all the topics of Introduction to Building at Stage I and Construction II at Stage II of the SCOTEC Certificate in Building.

Main references to building regulations within the text relate to The Building Regulations 1976 which do not apply in Scotland. Construction of buildings in Scotland is covered by the Building Standards (Scotland) (Consolidation) Regulations 1971–75. This fact is recognised in the text by the inclusion of asterisks which relate to equivalent sections of Scottish regulations listed in the Appendix. In addition, where Scottish practice differs significantly from English practice this is indicated.

It is hoped that the emphasis on practical and positive detailing will assist in providing the student coming fresh into the building industry from secondary education with a clear and positive identification with the essentials of good construction.

Module A First Fixing Joinery and Windows

1.00 First fixing joinery and its protection

Certain items of joinery must be fixed into the superstructure of a building before the finishing trades commence work. The reasons for this are as follows:

(i) Wet carcassing trades such as concreting, brick and blockwork by reason of the irregularities in the basic material or wide tolerances allowable in laying do not provide the degree of accuracy necessarily acceptable in the wholly finished building. These tolerances or unavoidable inaccuracies are taken up by the finishing trades of plastering and floor screeding, increased thickness (dubbing out) being provided to bring the surfaces level or vertical.

(ii) To effect this 'levelling up' process it is necessary for guides to be provided, carefully set, either horizontal or vertical as required, to enable the workmen keep their plaster or screeded surfaces to the required finished plane. In addition, these guides ensure that the thickness of plaster or screed is controlled to that specified.

1.01

There are two methods for controlling the thickness and plane of wet applied finishes:

(i) Softwood battens are fixed to the perimeter of openings in walls projecting sufficiently to control the thickness of the first (rendering) coat of plaster.

A more recent method is to fix expanded metal plastering beads which are left in place controlling not only the thickness of the plaster but also protecting the arris (corner edge) during the remainder of the contract.

(ii) Where the finish runs up to and against a permanent element of structure, e.g. a door lining, this can be so designed as to act as a ground, the plasterer using this to control the thickness and plane of his material.

1.02

Permanent elements of structure and fixed items which are used to control and enclose areas of wet wall finishes are:

(i) Sawn timber grounds fixed to wall surfaces at floor level to enable skirtings to be fixed at a later date.

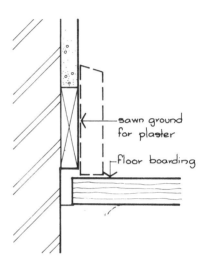

(ii) Sawn timber grounds fixed to the perimeter of openings to enable hardwood or expensive and fragile door frames and linings to be fixed later when the risk of damage by plasterers is over.

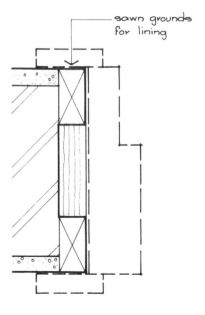

These grounds are nailed to the backing walls with cut steel nails driven either into the mortar joints of brickwork or random to lightweight blockwork.

(iii) Internal painted softwood door linings fixed upright and square to oversized openings left in brick or block walls, made to a width equal to the wall thickness plus the thickness of plaster on each or one side only as applicable.

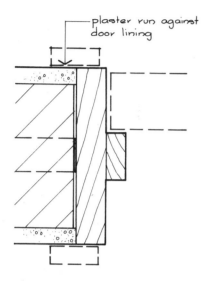

(iv) External door frames and sills provided with rebates to enable the plaster to be set into the frame to master the junction when drying shrinkage opens up the joints.

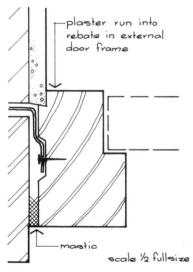

(v) External window frames (with or without the sashes or metal window frame inserts) with rebates to master the plaster junction as (iv) above.

(vi) Painted softwood window boards fixed to the internal sills of windows, cut within the width of the opening to enable the plaster to be run to their perimeter and avoid repairs or 'making out' later in the work.

(vii) Staircase flights whose side supporting members (strings) are normally grooved or rebated to receive the plaster as (iv) above. In addition, the perimeter of the stair well should be provided with the finished vertical board (apron) either grooved to receive the plaster or set with its bottom edge flush with the finished surface and the joint covered (mastered) by a cover fillet.

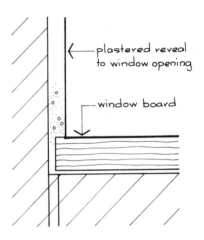

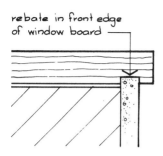

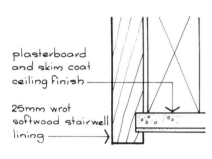

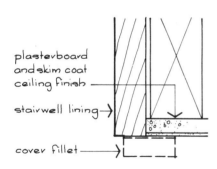

1.03

Floors are not 'first fixing joinery' as they are fixed early on in the contract, to enable internal access to be provided for all trades about the work. In most contracts concrete floors are cast as the superstructure reaches the appropriate level and timber or chipboard flooring is fixed and protected as soon as the roof is weatherproof.

1.04

Timber for joinery must be selected with great care. BS 1186 Pt. 1 *Quality of Timber in Joinery* gives good advice on selection. Certain timbers are more suitable for particular use or situations than others and the following is a good guide for selection:

External joinery	Hardwood	Softwood
Window sills	Afromosia, Agba, Guarea, Iroko, Keruing, Oak, Utile	Cedar (Western Red) Pitch pine,
Window and door frames		Cedar, Douglas fir, Redwood
Sashes and casements		Cedar, Douglas fir, Redwood
Doors		Cedar, Douglas fir, Hemlock, Yellow pine, Redwood.
Internal joinery		
Door frames and linings	Afromosia, Agba, Ash, Beech, Elm, Iroko, Meranti, Oak, Obeche, Ramin, Sapele, Utile	Cedar, Douglas fir, Hemlock, Parana pine, Pitch pine, yellow and white pine, Redwood, Whitewood.
Stairs	Afromosia, Agba, Ash, Beech, Gurea, Iroko, Mercanti, Oak, Sapele, Sycamore, Utile.	Cedar (yellow), Douglas fir, Hemlock, Parana pine, Redwood, Whitewood.

1.05

Much of the softwood used for joinery today is treated to protect it from decay or infestation. This treatment is especially important in respect of external joinery.

The timber must be specified, already treated from the suppliers. Usually treatment with a water-borne composition in accordance with BS 4072: 1974 *Wood Preservation by Water-borne Compositions* is required. Impregnation of timber is described in *Building Technology 1* (21.02). The degree of salt retention required for window and door frames to comply with BS 4072: 1974 is 5.3 kg/m^3. Cut faces should be dipped in preservative before the joint is made.

1.06

The moisture content (m/c) of timber used for joinery and the control of moisture content in the finished work is important. Most timber used for building joinery is kiln dried. When this timber is exposed again to the atmosphere it absorbs moisture and expands, continuing in this way until the surface of the timber is protected by priming or sealing against further absorbtion.

Timber for joinery, particularly internal joinery, should be used at a moisture content closely approximating to the average humidity likely to be experienced. Centrally heated buildings require joinery timber to be at a lower moisture content than if intermittent heating only is to be provided and the building should be dry and reasonably warm before second fixing joinery (doors, panelling and fittings) are installed. Failure to do so will produce swelling and sticking of doors and windows. Excessive heat, on the other hand, will produce excessive shrinkage, causing the cracking and splitting of joinery. Joinery with one face exposed to higher humidity e.g. external doors, is prepared from timber with a higher moisture content than wholly internal joinery. The following are satisfactory percentages for joinery timbers:

	Min m/c %	Max m/c %
Internal joinery (continuous heating)	10	12
Internal joinery (intermittent heating)	14	17
Internal doors (intermittent heating)	12	15
External joinery	17	20
External doors	15	18

1.07

Joinery timbers dried to the correct moisture content must be protected from absorbtion of moisture. The timber should be brought into the warm dry joiner's shop or store as soon as delivered and kept there until needed. As soon as the work is made up the finished joinery should be knotted (if required) and properly primed with a good quality wood primer and stored under cover until required on site.

Joinery should be protected from the weather during transit and on site should be stacked clear of the ground and well sheeted down to protect it from the weather.

1.08

The protection of joinery built into the superstructure of a building is required for the following reasons:

(i) To control the moisture content to a percentage suitable for the environment to be occupied (1.06).

(ii) To protect the joinery from damage caused by chemical reaction from materials forming the supporting walls. Examples of this are the rusting of ferrous metals (of which steel framed buildings with exposed sections are an example), efflorescence caused by leaching salts from brickwork and cement based blockwork and attack by sulphates.

(iii) To contain the exudation of oils and resinous matter from hardwoods and softwoods which can interfere with the hardening and adhesion of paint. Softwoods such as larch, spruce, Columbian and Oregan pine and hardwoods such as afromosia, agba, gurjun and keruing, all suffer from this problem and must be treated accordingly.

(iv) To protect and expose the natural colour and surface of hardwoods.

1.09

All joinery to be painted is first treated to inhibit exudation of resinous matter. With softwoods showing clear knots and pockets of resin on surface, these are treated by brushing on two coats of shellac knotting complying with BS 1336: 1971, knotting being a compound of shellac dissolved in methylated spirit. This seals the knots and resinous areas.

Softwoods and hardwoods which are to be painted, and at the same time are generally resinous, need a primer with greater inhibiting powers than that provided by the normal wood primer. In these cases an aluminium-based primer should be applied. Otherwise timbers should be primed with a good quality commercial wood primer.

Where hardwoods or softwoods are left to show their natural colour and grain, the timber must be treated before leaving the joiner's shop with a coat of varnish or polyurethane lacquer to seal the surface against moisture. In addition, the bedding faces of the joinery should be primed, preferably with an aluminium based primer to seal the surface against both chemical action from the wall and moisture.

Metal window sections are usually either constructed from galvanised rolled steel sections or formed from extruded aluminium sections usually protected by anodising. While aluminium windows are generally left unpainted, steel windows being protected by their galvanised coating from rusting merely need treatment to provide a base for the application of usual gloss paint systems. This is usually obtained by painting the metal surfaces, all faces, before fixing, with a calcium plumbate primer which bonds well onto the galvanised surface.

1.10

Joinery is often made with horizontal projections (horns) to head and sill to assist in stabilising the installed unit in the wall. Often these horns are cut and adapted on site to suit the bond of the wall. The exposed surface of the timber must be re-primed before the walling covers the exposed part or rot will certainly commence in the bare wood.

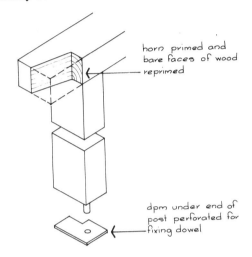

A similar problem can occur in door frames where the ends of the side posts are set down onto a brick or concrete sill. The bottom is exposed to excessive moisture in the end grain even when well primed and is best protected by a strip of polymer or Code 4. lead cut to profile and set under the primed end of the post.

1.11

Where the side members of external joinery meet the reveal of cavity walling it is an advantage to either:

(i) Set the d.p.c. into the plaster rebate provided to prevent moisture passing the gap and showing on the plaster face, or

(ii) setting the d.p.c. deep into the back of the metal window section for the same reason, or

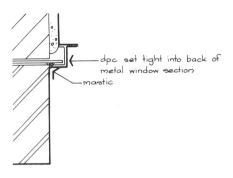

(iii) tacking the d.p.c. to the bearing surface of the frame and bending it to pass through the gap between the cavity closer and the outer leaf of the wall (225 mm wide d.p.c. is needed here).

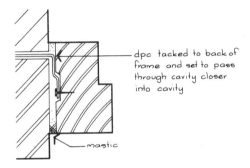

Window sills should be set down on a strip of d.p.c. to provide extra protection at this point.

The gap between the frame and the outer leaf of the wall should be sealed in mastic in preference to the mortar which tends to break up and fall out within a short period of time (1.02(iv)).

1.12

First fixing joinery must be protected from physical damage after fixing for the remainder of the contract.

Physical damage can be caused by barrows of mortar for wall rendering coats and careless movement of material, etc. Painted door frames and linings are usually protected by nailing waste ply and hardboard strips to the lower portions to take abrasions and knocks. Window boards are protected by similar means, care being taken to protect the edge of the board from damage. Hardwood which is to be exposed is often either cased in plywood or wrapped in heavy gauge polythene.

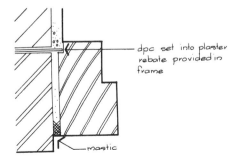

Nail holes and marks on painted joinery are filled and made good before further paint coats are applied.

2.00 Windows and window boards

Windows for small domestic structures are generally made from one of three materials, wood, steel or aluminium. Each material has its own particular problems, advantages and uses, and each can be either supplied as a standard stock item or purpose made for the particular contract.

Standardisation is generally employed for metal units due to the cost of special jigs but timber windows can be procured to special sizes and designs at very little extra cost to standard items.

2.01

The construction of windows and, in fact all joinery, is today largely dependant on the use of specialised woodworking machinery which is set up by the machinist to carry out the precise and particular cutting labours for which it was designed.

Timber is generally supplied in standard scantlings and lengths and, after selection, the timber is sawn to length on a cross-cut saw. The piece of timber is then cut to its approximate width on a circular saw. This saw is provided with a flat table which can be tilted to an angle to enable bevel cutting to be carried out.

The roughly sized timber is then passed through a planing machine which not only planes the faces smooth but also ensures that the angles are precisely 90°. Passing the planed section through a machine called a thicknesser allows the operator to reduce the wood to exact dimensions.

There are several machines available for producing moulded surfaces on timber. Some machines are provided with cutting blocks each capable of cutting the timber in turn to progressively enlarge and develop a particular moulding. A simple machine much in use is a spindle moulder which forms mouldings on straight, curved or irregularly shaped pieces of timber.

Mortices are required to accommodate protruding tenons used for jointing timbers at right angles and these are prepared by a mortising machine incorporating either a hollow chisel with a central revolving bit or a chain cutter. The complementary machine to this is the tenoning machine which produces either single or double tenon depending on the way it is set up and the type of cutter block used.

Finishing of joinery is either by hand, where involved or small section work is concerned or by drum, belt or disc sanders operating over a table on which the work may be rested.

2.02

Softwood is specified to nominal sawn sizes and, as will be appreciated, the action of smoothing (preparing) the surface by either hand or machine planing reduces these nominal sizes. CP 112 *The structural use of timber* requires that the surfacing reduction on both American and European stock softwood shall not exceed the following:

Nominal dimension (mm)	Reduction (mm)
25.6 to 76	3.0
76 to 305	6.5

When detailing joinery, these reductions in nominal dimensions must be taken into account.

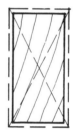

Diagram to show surface reduction from sawn to prepared timber

2.03

Steel windows for domestic purposes are made from rolled mild steel bars and sections in accordance with BS 990 Pt. 2: 1972. After cutting and welding to form the required window unit the sections are cleaned, to remove all dirt and grease, and then immersed in a bath of molten zinc to provide a durable rust resisting finish.

In low cost work, steel windows are fixed by lugs directly into the prepared openings but in better class work, the windows are first screwed into prepared softwood frames which are fixed in the same way as timber windows. The joints between the steel and the timber frame are pointed up with mastic to exclude water penetration.

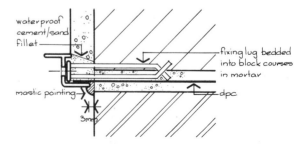

Fixing standard steel window section direct to cavity wall

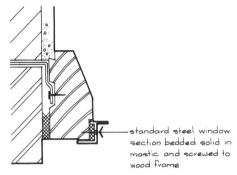

Fixing metal window section into prepared wood frame

2.04

Aluminium windows for domestic purposes are of different types, e.g. pivotted, sliding and casement, all manufactured from extruded sections jointed either by corner cleats or direct screw fixings and to comply with BS 4873: 1972 *Aluminium alloy windows*. The metal is protected from corrosion by anodising the surface either a clear natural or a coloured finish.

Aluminium windows are generally fixed directly into prepared openings in the wall.

2.05

Windows are required for two principal purposes:

(i) To provide natural daylighting into the building, and

(ii) to provide natural ventilation into rooms where this is essential to the health and welfare of the inhabitants and where fresh air cannot be provided by mechanical means.

In addition windows provide the occupants with views of the external surroundings, gardens, streets and the countryside, improving the environment by opening up the room to the outside world.

2.06*

The Building Regulations Section K require precise minimum requirements for the ventilation of habitable rooms, i.e. rooms such as living rooms, bedrooms, kitchens and sculleries, as follows:

(i) The total area of ventilator(s) must exceed 1/20 of the floor area of the room, and

(ii) part of this ventilated area must exceed 1.75 m above the floor (K4).

In addition, regulations require that larders where perishable foods are stored must be provided with either

(i) a window with a ventilated opening at least 85 000 mm^2 protected by a durable fly screen, or

(ii) two closeable ventilators, one in the upper half and one in the lower half of the external wall having an unobstructed area of at least 4500 mm^2 and fitted with a durable flyscreen (K6).

2.07*

The area of window required to be provided for habitable rooms must not be less than 1/10 the floor area and the minimum floor to ceiling height shall not be less than 2.3 m except under beams and in bay windows where the clear height may be reduced to 2.0 m (K8).

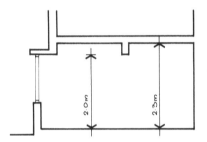

2.08

A number of types of opening windows (lights) have evolved over the years to cater for the differing needs, applications and fashions prevailing at the time. Not all of these have been entirely successful, many being limited by the available level of technology and also cost.

Most types are available in timber or metal construction and can be summarised as follows:

(i) Top hung ventilators with restricted opening give good security and are reasonably weatherproof (timber/steel). These opening lights are hung on either butt or cranked hinges (when stormproof timber window sections are employed) and secured to the bottom of the frame with a short (203 mm) casement stay and pin.

(ii) Side hung casement lights provide a good opening range from minimum to maximum 90° opening (timber/steel/aluminium). These opening lights are hung on hinges as the ventilators and secured to the frame by means of a casement fastener, and the sill by means of a long (305 mm) casement stay and pins.

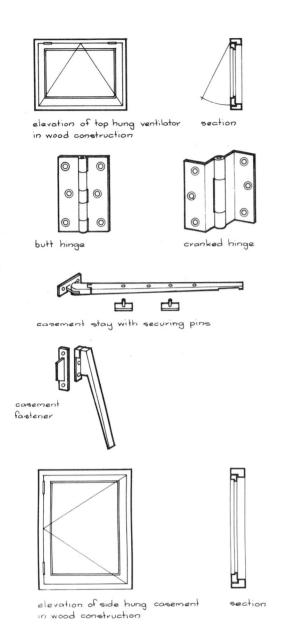

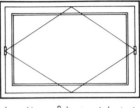

reasonable wind pressure at the opening angle set by the operator. Vertically-pivotted windows are secured by casement fasteners to the side framing. Horizontally-pivotted windows are secured either by a sprung fanlight catch at the head or a mortice cockspur fastener at sill level.

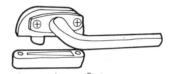

(iv) Sliding sash windows which can be either:

(a) Vertically sliding, when the weight of the glazed sash (light) is carried by either weights or spiral

(iii) Pivotted lights which are of two types:

(a) Vertically-pivotted in which the pivots may either be central (steel/aluminium) or set towards one edge (steel).
(b) Horizontally-pivotted in the centre of the light (timber/steel/aluminium)

Pivotted timber windows are provided with metal back flap hinges which are let into the frame and the side of the pivotted light. Metal windows are provided with metal pivots which form a part of the construction. These pivots are so made that they operate under friction to hold the window in position against

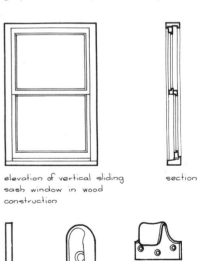

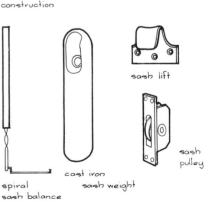

sprung balances in the box side framing (timber/ aluminium), or

(b) Horizontal sliding, when the weight of the sash is carried on runners of metal or nylon (timber/ aluminium). These windows are provided with sash lifts and sash locking screws for vertical sliders and finger pulls and locks for horizontal sliding windows.

(v) Hopper windows, hinged on the bottom and opening inwards at the top with side cheeks of pressed steel or quadrant stays to restrict the opening (steel/aluminium). These windows are secured closed by means of sprung fanlight catches.

2.09*

Timber windows are in general use for both standard and special situations. Standard windows in a range of restricted and preferred sizes are relatively cheap and effective in preventing weather penetration. Where special sizes and types are required, timber is the cheapest and most effectual method when numbers are restricted.

In addition to providing natural daylighting and ventilation a window must be weathertight in the exposure to which it is subjected and must continue to operate satisfactorily and without undue maintenance for the expected life of the building. Consequently the basic material must be sound and of good quality and, in view of its exposure to the weather, preferably treated before manufacture to protect it from decay (1.05).

To achieve satisfactory operation and to ensure minimum maintenance and weather-tightness, timber windows should be designed to expedite the passage of water over their external surface and prohibit water penetration through gaps around opening lights. To this end horizontal surfaces are avoided, sloping (weathered) surfaces being provided and grooves (checks) provided in the spaces between opening lights and frames to stop capillary attraction of rainwater at these points.

The details in the adjoining illustrations show good practice in the design and detailing of side hung sashes in timber windows.

2.10

A variation of this detail is provided by the stormproof window developed by the English Joinery Manufacturers Association (EJMA) which provides a cover mould to protect the open joint between opening light and frame and a small hood at the head to assist in throwing water draining off the structure clear of the window.

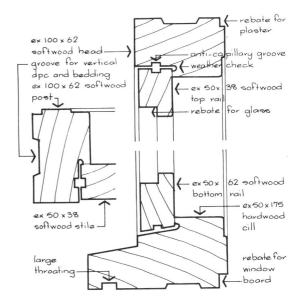

Plan and section of wood casement frame and opening light

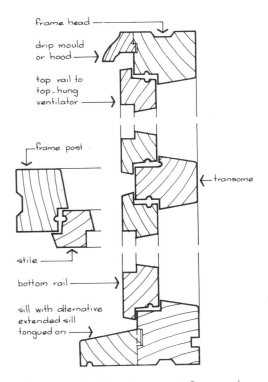

Plan and section of softwood casement frame and opening light to BS 644 Part 1

2.11

Vertically sliding sash timber windows are virtually weatherproof by reason of their design and allow for the use of large sheets of glass in their glazing. More expensive than simple casement windows, they are usually made to order for particular contracts.

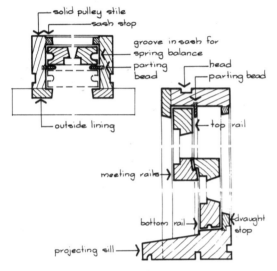

Plan and section of softwood double hung sash window (spring balance type) to BS 644 Pt. 2

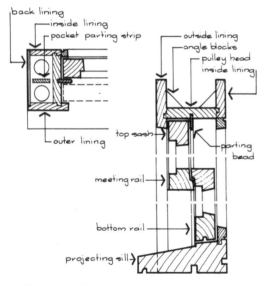

Plan and section of softwood double hung sash window (weight balanced type) to BS 644 Pt. 2

2.12

Steel windows are fabricated from standard rolled sections either to standard sizes and assemblies or specially to order. In most cases, the steel sections are supplied fixed into standard timber subframes and these are built into the structure in the same way as a standard timber window unit (2.03).

Where the steel sections are built direct into the structure care must be taken to ensure that the weathering d.p.c. is properly set into the metal section and that this is pointed up in waterproof cement and soundly pointed up round in mastic (1.11).

2.13

Aluminium windows are generally provided with wide extruded sections which are fixed, weathered and pointed in a similar manner to steel windows when these are not provided with a timber sub-frame.

2.14

The interior horizontal surface behind the bottom member (sill) of the window frame is usually protected and finished by one of a number of alternatives. The selection of the right material will take into account one or other of the following considerations:

(i) The use or misuse likely to occur during the life of the building, e.g. unsuitable, abrasive or wet objects placed on the surface.

(ii) The decorative effect required.

(iii) The quality of the work or building.

2.15

A number of alternative materials are in general use and these are as follows:

(i) Softwood, at least 25 mm nominal thickness, the boards cross tongued if the width exceeds 225 mm, the front edge projecting beyond the face of the plaster by about 19 mm. The back edge of the board should be set in a rebate in the sill to avoid shrinkage cracks which will open up as the board takes up its final moisture content (shrinkage across the grain). Softwood window boards are always primed all faces before fixing and generally finished with a gloss paint finish (1.02(vi)).

(ii) Hardwood boards of similar thickness and construction to softwood. The free edge is often moulded and provided with a small scotia in good quality work. The finished surface is usually scraped down and sealed with a clear matt natural finish such as matt polyurethane lacquer with the bedding face primed (1.09).

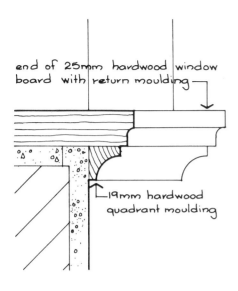

Wood window boards are either secured to the wall with galvanised steel wire or straps or screwed down to pallets (wood blocks) built into the vertical mortar joints, the screw heads covered with wood discs set flush with the exposed surface (pelleting).

(iii) Quarry tiles of various thickness and size from 152 x 152 x 15 mm to 230 x 230 x 24 mm. The smaller tiles are finished on the free edge with a course of round edge tiles (bullnose) but larger tiles are usually set to project the square edge about 29 mm clear of the plaster face. The tiles are bedded in cement and sand (1:4) and pointed up in white cement grout.

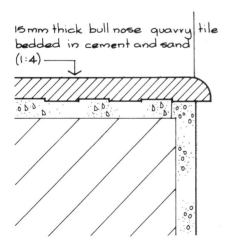

(iv) Glazed wall tiles to match those fixed to the walls (kitchens and bathrooms) bedded in adhesive onto a flushing up of cement and sand (1:3) applied over the horizontal area, and grouted up with a proprietory grout to seal the joints. Free edges are finished with a bullnose tile where available.

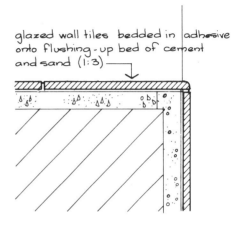

3.00 Door frames and linings*

The need to provide a permanent stop for wet internal wall finishes has been fully discussed (1.00). Timber door frames and internal linings provide, as a first fixing, such a stop. In addition both fulfill other functions in the permanent enclosure and security of the building. These may be summarised as follows:

(i) They regularise the precise perimeter of the opening ensuring that a pre-determined enclosure element (door) will fit the opening subject to the employment of specific tolerances.

(ii) They provide a secure and reasonably unrestricted fixing background for hinges and simple housings for locks and other security ironmongery.

(iii) They provide a weatherproof seal to the gap between the door and the superstructure.

3.01

The construction and sizing of door frame and lining members follow traditional experience. Obviously the size of the members will depend on the size of the opening and the weight of the doors to be fixed to the side members of the frame (posts or jambs). With very large or heavy doors it may be necessary to hang these from hinges carried by fixings direct into the super-structure avoiding placing any load on the frame or to provide overhead track and bottom channels and guides to transfer the load to the floor and beam supporting the opening. In most single and double leaf doors not exceeding 900/1000 mm in width and a height of 1950/2000 mm frames carry the weight of the doors on traditional metal hinges.

Hinges are usually secured to doors and frames by screws, generally of steel but matching hinges when these are of brass or bronze and fitted to hardwood doors. The screws are of stout gauge and generally 25/32 mm long, depending on the load. Frames and hinges must be thick enough to ensure that the screws, when driven home, are fully contained within the thickness of the wood. In general, a frame should be made from timber at least 65 mm thick and a lining from 32 mm minimum thickness of board.

3.02

The strength of a frame or lining lies mainly in the manner of its fixing and in the manner of its jointing.

Door frames are generally of timber sizes 125 x 65 mm thick and the joint between the side post and the head is usually made by forming a tenon on the end of the post and securing this into a mortice cut in the head member with glued wedges and a hardwood dowel. The free end of the head (horn) is splayed back for building into the wall to provide extra security. A similar joint secures the sill.

Door linings being constructed of lighter material and carrying lighter doors need only to be tongued together at the head. To ensure rigidity in transport and fixing a temporary batten is splay fixed to the foot of the lining and a diagonal brace to one top corner. No sill is provided but often a hardwood threshold is provided against which the carpet or floor finish can be stopped.

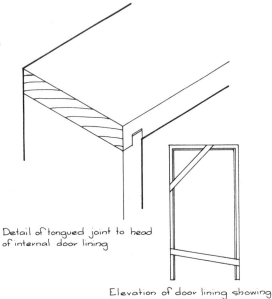

Detail of tongued joint to head of internal door lining

Elevation of door lining showing temporary bracing

3.03

The problem of excluding weather around external doors lies primarily with the design of the frame and the provision of weather checks. Care must be taken to prevent penetration by capillary attraction—checks should be provided both to the edge of the door and in the frame profile to post and head. These must be deep and wide enough to be effective.

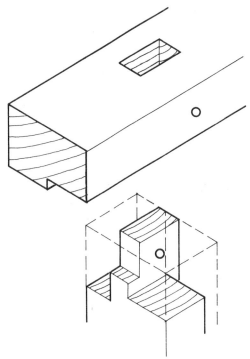

Detail of mortice and tenon joint at head of door frame

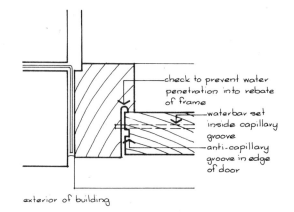

exterior of building

check to prevent water penetration into rebate of frame
waterbar set inside capillary groove
anti-capillary groove in edge of door

3.04

The threshold poses special problems. Here gathers an accumulation of storm water, running down the impervious face of the door and forced under it by wind pressure. A good weathered slope to the threshold must be provided to help this accumulation to run off and a high galvanised iron water bar provided to the threshold. Extra protection is gained from an oversailing weatherboard with undercut bottom edge. As water tends to penetrate around this feature near the frame posts care must be taken to place and extend the water bar to prevent water penetrating to the interior of the building.

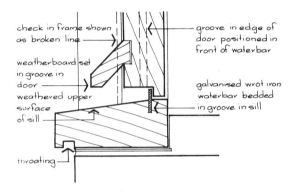

3.05

The frame section at the head follows the profile of the posts so long as the frame is set in the rebate to obtain protection from the supporting wall overhead. Otherwise a weather bead should be provided and tongued onto the head to throw stormwater down off the top joint.

Lead trays, when provided, should be dressed down the face of the head at least 25 mm and cut off straight.

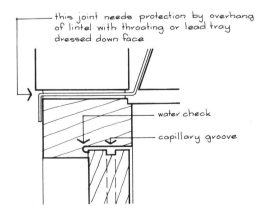

3.06

Internal door linings are generally provided for the full thickness of the wall plus the plaster. Most internal walls are of block or half brick in thickness and thus there is little problem in the sizing of the member.

Two methods of forming linings are in general use:

(i) By using timber 38 mm thick and forming a rebate 12 mm deep and the width of the door +3 mm to accommodate the door itself. This method is commonly used for heavy internal doors or where the lining is fixed to a thin wall of lightweight blocks (say 75 mm thick). With lightweight block partitions, the lining is formed the full height of the room plus 150 mm, cut back above the head to the thickness of the blockwork and fixed to the floor above. This ensures greater rigidity not only to the block wall but also to the lining. The space over the head is filled in with blockwork and the timber covered with expanded metal lath to ensure support for the plaster. These linings are called 'storey frames', and can, as an alternative be prepared for glazing to form fanlights or borrowed lights.

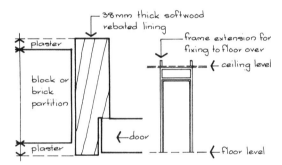

Detail of softwood rebated lining and storey frame

(ii) By fabricating the lining from 32 mm softwood and forming the rebate for the door by nailing (planting) on a softwood bead (stop) size 38 x 19 mm to suit the door thickness.

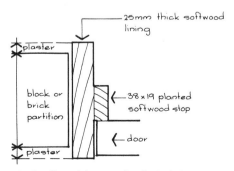

Detail of softwood lining with planted stop

3.07

Internal door frames for fire check doors are required to be constructed to give a notional degree of fire resistance. Frames are usually fabricated from 65 mm thick softwood and provided with a rebate 25 mm deep to accommodate the 45 mm minimum thickness of door (22.03 and 22.04).

3.08

External door frames are fixed by building in the horns at threshold and head or, if no threshold is provided, by inserting galvanised tube into the end of the frame and setting this in a mortice formed in the threshold and running this solid with neat cement grout.

Internal door frames and linings are either screwed to hardwood plugs (pellets) built into the brick joints to the reveals or provided with galvanised fixing cramps screwed to the back of the frame every 450 mm to suit brick or block joints, the ends being set in mortar. Where stud partitions are employed, the linings are nailed to the timber studs forming the opening trimming.

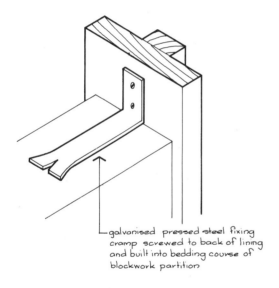

galvanised pressed steel fixing cramp screwed to back of lining and built into bedding course of blockwork partition

4.00 Staircases and landings

A stair is a continuous run of steps leading from one floor or level to another. Each continuous series of steps is called a flight and horizontal areas between floors are called landings. The horizontal portion of a stair is called a tread and the vertical a riser. Both are generally housed (let into) side members which are known as strings but there are variations to this method of construction.

Protection afforded to the open side of a stair is known as a balustrade, being either solid or filled in with vertical rods known as balusters and crowned with a handrail.

4.01*

The Building Regulations (Section H) lay down stringent requirements for domestic staircases. These may be summarised as follows (H3):

(i) All treads shall be parallel (or tapered) and level.

(ii) All risers to be uniform throughout the length of the flight and equal in height.

(iii) The width of the tread must not be less than the going at any point.

(iv) The length of the tread must not be less than the width of the flight.

(v) All treads must have the same going.

(vi) The nosing of open treads must overlap the tread below by not less than 15 mm and the gap between the two not exceeding 100 mm.

(vii) Tapered treads must have the same going at the centre, the same rate of taper, the narrow ends must all be at the same side and not less than 75 mm wide.

4.02*

The Regulations lay down four building purpose groups for which the requirements of staircases vary, domestic buildings being Groups 1 and 3 (Residential), single dwellings in which the staircase is for its exclusive use being Purpose Group 1. The detailed requirements of Regulation H3 to this group are as follows:

Width of staircase	800 mm. minimum (generally)
Pitch of flight (i.e. angle of staircase)	42° maximum
Number of risers per flight	2 (min.) to 16 (max.)
Height of riser	75 mm. (min.) to 220 mm. (max.)
Going of tread	220 mm. (minimum)

The relationship of riser to going is $2R + G = 550$ mm(min)/ 700 mm (max.).

A handrail must be provided for the whole length of a flight less than 1 m wide at a height between 840 mm and 1 m measured vertically above the pitch line. Headroom above the staircase must not be less than 2.0 m measured vertically above the pitch line. Clearance must be not less than 1.5 m at right angles to the pitch line.

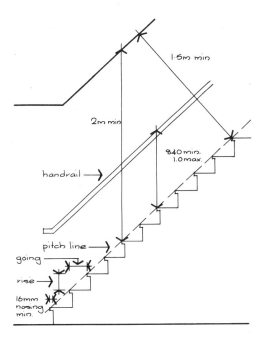

(a) by a half space landing which reverses its direction through 180 degrees

(b) by a quarter space landing which reverses its direction through 90 degrees, or

(c) by a quarter or half space landing formed from winders also reversing its direction through 90 degrees.

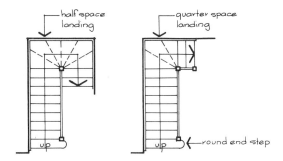

possible winders shown as broken line

(iii) The *Geometrical stair* in which the strings and often the handrails are continuous, often circular on plan with the steps radiating from the central open well.

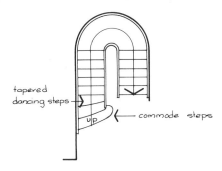

Variations and combinations of stair types can be produced for special circumstances.

4.03

In small domestic buildings space is often very much at a premium. Consequently, in the past, a great variety of staircase layouts have been devised to fit vertical access into the smallest plan area. Many of these ingenious constructions are now too costly for modern dwellings and the construction of staircases has become limited by economic factors. The following are in standard use today:

(i) The *Straight Flight stair* mounting in one direction only throughout its length. It may consist of a single flight only contained

(a) between two walls when it is known as a cottage stair, or

(b) set against one wall with one side open to the ground floor hall

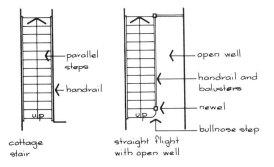

(ii) or it may consist of two or more flights in its length separated

4.04

Openings in timber floors have to be specially formed to accommodate staircase access. The ends and weight of joists need to be supported and the total load transferred onto the walls. The opening is formed by providing a trimmer joist at right angles to the span usually the same depth as the rest of the joists but 25 mm thicker which is set parallel with the long side of the opening or across the head. The ends of the joists are then connected to this trimmer by either:

(i) a traditional housed joist, or

(ii) galvanised mild steel plate connectors nailed to both trimmer and trimmed joists.

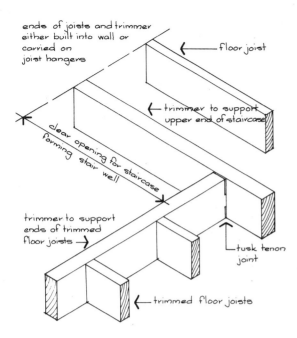

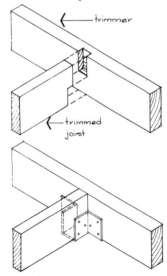

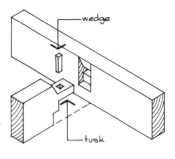

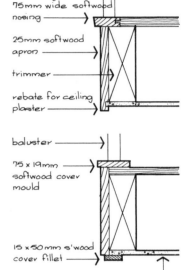

Alternative details of apron lining to staircase well

It is sometimes necessary to provide two trimmers to an opening and to ensure their joint stability the connection is made by a tusk tenon joint which locks the timbers together.

The free edges of the trimmed opening are faced up with 25 mm softwood aprons surmounted by either a nosing of similar thickness to the floor boarding or by a cover mould to master the end grain on the cut ends of the boards.

4.05
The construction of a traditional timber staircase of types (i) and (ii) (4.03) follows simple well defined principles. The nominal thickness of the tread should not be less than 25 mm (32 mm preferred) and the riser less than 19 mm. The riser should be tongued to the tread as shown on the detail and screwed together with long steel screws. Sawn softwood blocks are glued to the angles between tread and riser to increase rigidity and prevent 'cracking'. The nosing should project the thickness of the tread and be either slightly rounded, splayed and rounded or, with stairs which are to be left natural finish a small scotia mould added.

4.06
Support for treads and risers is provided by strings* formed from wide boards 32 mm thick. These may be either plugged and screwed to the wall, or left open

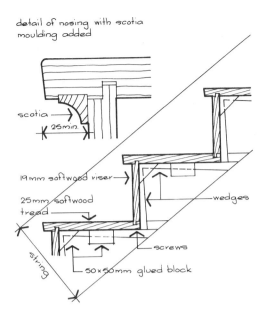

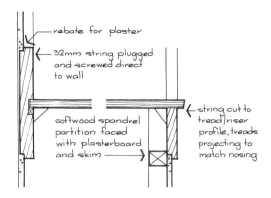

to the well when they are provided with either a capping when used as a base for balusters or provide the support for a solid panelled balustrade.

The space (spandrel) under the latter type of stair can either be filled in with blockwork plastered on both sides or timber studding finished with plaster board and skim coat plaster.

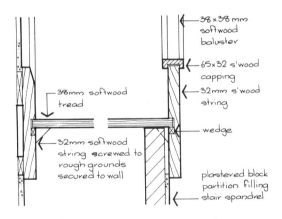

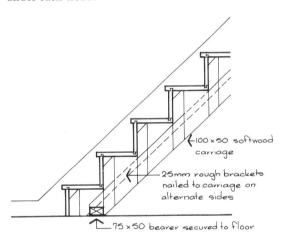

carriage piece running down the centre of the stairs and supporting the underside of the internal angle between tread and riser. Rough brackets 25 mm thick are fixed on alternate sides to provide support under each tread.

The carriage is birds-mouthed over a batten secured to the floor at the foot of the stairs and secured in a similar way to the trimmer at the head.

4.07

Open balustrades are formed with square or turned vertical rods not less than 32 mm thick, set vertically not exceeding 100 mm apart. It is usual to provide a capping to the string to which the balusters are dowelled or housed. The balusters are similarly dowelled into the underside of the handrail.

Support for the handrail at both ends is provided by a newel post usually from 100 x 100 mm prepared softwood and with a moulded capping as a terminal. The open string is secured at top and bottom of the flight to the newels by shouldered double tenons secured with dowels.

An alternative to the open, is a solid balustrade which can either be constructed by setting over the

An alternative string detail cuts the top of this member to profile following the line of tread and riser, the nosing of the former being returned across the external face of string to the bottom of the upper riser.

This is known as a cut string detail. The ends of the treads and risers are housed into tapered grooves (housings) cut into the strings and secured with glue and long tapered wedges, extra rigidity being provided by small triangular glue blocks.

In stairs over 900 mm wide it is usual to provide an additional timber member called a bearer or

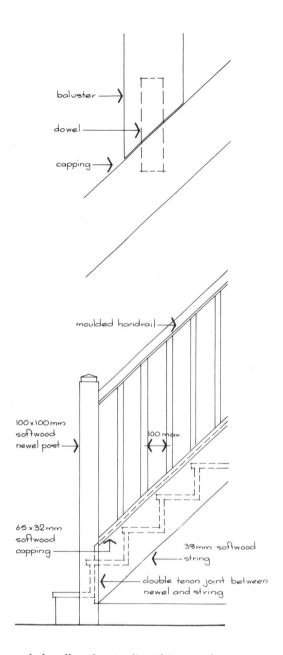

4.08

Winders are sometimes necessary where space is limited. In practice, only three risers can be accommodated in a quarter turn due to the limitations of the Building Regulations.

To construct the winders, 50 x 25 mm sawn bearers are set across in positions directly under the riser from the newel to the wall string to provide additional support necessary due to the additional width of the winder treads. The treads and risers are otherwise constructed and housed both together and to the string in a similar manner to the straight flight.

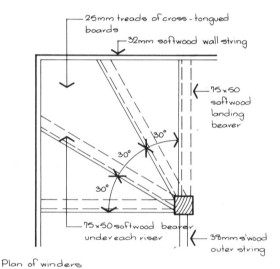

The maximum width of window is 75 mm

4.09

Half and quarter landings are constructed in a similar manner to floors. The end of the straight flight is carried on a 175 x 75 mm trimmer and the landing filled in with 100 x 50 mm sawn joists at 400 mm centres to carry the boarded floor. The line of the string is continued by making out around the perimeter of the landing in matching skirting.

spandrel wall and extending this past the string, terminating in a wooden handrail capping, or by constructing a sawn softwood framing faced both sides with plywood or a similar sheet material and providing a capping in a similar manner to the detail above. Care must be taken to securely fix the free end to prevent movement.

4.10

Geometrical staircases in timber are usually made in laminated timber by specialist joinery contractors. In most cases the external finish is in hardwood either solid or applied as a veneer.

Module B Services and Drainage

5.00 Services

To produce a satisfactory environment for modern standards of living, the installation of services has grown from the 19th century standard of a simple brass water tap provided over a brownware sink to a sophisticated and highly technical installation whose cost often equals or exceeds that allocated to the building structure itself.

Services can be grouped under three main headings:

(i) Services which contribute to a more convenient way of life and include electric lighting and power, heating and ventilation and air conditioning.

(ii) Services which contribute to a more efficient and healthy environmental unit such as sanitation and water supply installations.

(iii) Services which materially assist in the usage of the building such as lift installations and refuse disposal systems.

Certain services are basic to all buildings and include drainage and sanitation, heating, water services and electric lighting and power. Others are more readily associated with particular building types and usage.

Generally, services conform to certain basic principles, whatever the size of the installation; excessive demands being met by increased unit sizes and more advanced distribution technology. The services described in general terms in this module refer to small domestic buildings and incorporate the general principles and statutory limitations imposed on installations and systems of this size.

6.00 Public utility services and intakes

Certain utility services are regarded as basic requirements to modern living in most areas of Britain. These services are:

(i) Water.

(ii) Electricity.

(iii) Gas.

(iv) Sewage removal and disposal.

While all these can be expected to be available in urban areas, this is rarely so in the country where the cost of providing main supplies may be totally uneconomic when related to financial returns. However, in most places today, electricity is freely available and in all but remote country areas a supply of clean wholesome water is provided through the agency of the local water boards.

Gas is somewhat different, its distribution requires pipes of much greater size than water and consequently the capital cost for each connection is much greater. Gas is rarely available from the mains outside the areas of reasonably dense population.

6.01 Water supplies

An adequate supply of clean wholesome water is an essential requirement for all buildings, used for personal hygiene, drinking and cooking or for manufacturing processes.

Where a piped supply is not available, water is obtained from wells or springs and its suitability must be tested at intervals to ensure freedom from bacteria.

6.02

Water, whether obtained from wells or springs or supplied through the main, originates in condensation in the atmosphere and falls to earth in the form of precipitation, e.g. rain, hail or snow. As either surface water (streams, rivers or lakes) or ground water percolating through the subsoil until it reaches an impervious layer such as clay, it is collected and processed by the water supply authority. This processing comprises screening, sedimentation, filtration, chlorination, aeration and, in some areas, fluoridation, to ensure the water is fit for human consumption. It is then pumped into a closed reservoir for gravity distribution by mains for use by the consumers.

6.03

Two principal problems affect water mains laid in the ground:

(i) *Pressure from road traffic.* Most mains are laid under road carriageways as being the most direct routes to premises requiring water supplies. The increasing weight of road vehicles and the cost both of road construction strong enough to carry the wheel loads and of breaking up the construction and renewing it after main renewal or extensions are considerable items of expenditure.

(ii) *Frost action* which in severe winter conditions may cause water in pipes laid at shallow depths to freeze, with the likelihood of eventual splitting or fracturing of the pipe.

It is therefore a general rule that water mains are set at a sufficient depth below the surface to protect them both from pressure from above and the possible effect of extreme cold.

Water supply authorities' mains are generally laid out in the form of a circuit from which a grid of subsidiary distribution mains provide water for specific areas or districts. The materials used for these mains are either cast iron or asbestos cement, both materials being easy to tap under pressure for new connections.

6.04

The local water authority has a statutory duty (subject to certain conditions and reservations which do not concern the subject matter of this book) to afford a supply of mains water to a premises on demand.

Where the main supply is located close to the boundary of the site this provides few problems apart from the actual physical connection, but where the supply is at some distance, legal permission by means of wayleaves may be necessary if the connection or extension of the main supply has to be laid across land in the ownership of a third party. Powers are vested in the supply authorities enabling them to achieve this but it all takes time and the cost to the applicant may be expensive.

6.05

Connection to the supply main is effected by drilling and inserting a metal plug cock in the crown of the main pipe. From this a communication pipe is taken to a stop valve and protection chamber situated just outside the boundary of the site. A bend, known as a goose neck, is incorporated in the communication pipe to relieve any stress likely on the connection. The cost of all this is charged to the building owner.

The size of the communicating pipe will depend on the building to be served. For single domestic buildings this will be 15 mm copper pipe but the size may be increased if mains pressure is low.

6.06

Extension of the supply from the stop valve will be the responsibility of the contractor. Depth is as important as for the main and the pipe should be laid at a depth of at least 760 mm below the finished ground level. Where the pipe enters the building it should be protected by being drawn through a protective duct rising out of the floor at least 750 mm from the internal face of any external wall to protect the pipe from frost.

6.07

Suitable materials for supply pipes are PVC (BS 3505), polyethylene (BS 1972) and copper (BS 2871). Copper pipes laid in the ground should be protected by wrapping in a proprietary material to prevent mechanical damage during backfilling the trench and, in some soils, chemical attack. Plastic pipes are not affected by damage or chemical action.

6.08

Immediately on entering the building, the water main should be provided with a screw-down brass stopcock (BS 1010), preferably incorporating a drain cock to allow the main distribution riser to be drained down if required. Alternatively a separate drain cock can be fitted immediately above the stopcock.

6.09 Electricity supply

Electrical supplies are brought into a building in one of two ways, either

(i) by overhead supply from an Electricity Board supply pole. This is a common method in rural areas. The cables are supported on porcelain insulators fixed by means of a metal bracket to a high point of the building and from here the cables may be brought direct into the building through conduits (pipes for conducting electrical cables) to the meter position; or

(ii) by underground supply situated in the roadway to which is connected a service cable run underground into the building to the meter position.

Whichever method is used, the Electricity Board will carry out all the work, including making the supply

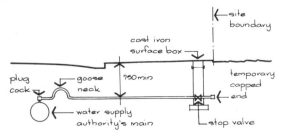

cable connection to the main, excavating the trench up to the building, for the cable and filling it in afterwards, drawing the cable into the building and making connection to the installation as discussed later.

6.10

As with the water main, it is necessary to ensure that, if required, the service cable can be renewed without having to break out the foundations. Underground armoured cables are about 30 mm diameter and are difficult to bend.

In practice a 100 mm stoneware or pitch fibre bend should be built into and through the foundations and floor slab terminating under the meter position, through which the cable can be drawn.

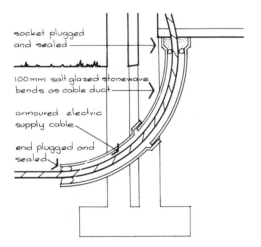

6.11

Overhead supplies are carried using separate cables for each phase, the cable being encased in an insulator protected by a fabric casing. The cable is known as HSOS.

These cables are relatively flexible and can be accommodated in 19 mm conduits run in parallel to the meter position with easy radius bends to facilitate drawing in the cables.

6.12

The supply cables, whether overhead or underground, are connected by the Electricity Board to a sealing chamber which incorporates a fuse to isolate the installation from any fault in the supply. From terminals provided in the sealing chamber, the Board's staff will connect short lengths of cable to their meter fixed alongside. Connection is made from the meter to the consumer unit which incorporates fuses controlling the circuits forming the electrical installation of the building.

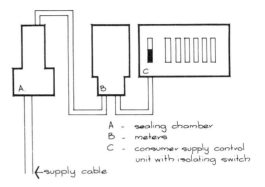

6.13 Gas supply

Gas supplies for cooking, heating and water heating mainly utilise natural gas supplied through a network of trunk and distribution mains, generally situated in the road or footway of public thoroughfares.

Connection to the main and the provision and installation of the service pipe into the building is carried out by the Gas Board, the service pipe being left 'capped off' and sealed about 200 mm above the floor in the meter position until required. As it is unlikely that the wrapped black steel pipe and fittings used for the gas service will require replacement during the life of the building no intake service duct is usually required, the pipe being merely concreted into the floor.

6.14 Sewage

The basic requirements of all drainage is to collect and discharge soil and/or stormwater to a public sewer without causing nuisance or hazard to health. The system must be self-cleansing and operate with the minimum of maintenance.

6.15

Public sewers operate on the water carriage principle; the water discharged into them from the separate connections being used to flush out solids and keep these moving in suspension.

The connection of private drains to public sewers is carried out by either the staff of the local authority or the water board. The usual method is to cut a hole in the top quarter of the sewer pipe and attach to it a saddle branch set at the appropriate angle and position to receive the drain.

As sewers are usually at a greater depth than other services due to the need to construct them to even and regular falls towards the outlet to facilitate their self-cleansing action, the drain from the last manhole to the sewer connection will often be at a steeper incline than that permitted for private drains.

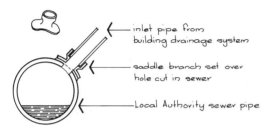

6.16

Where sewer connections occur under public highways, it is often a requirement of the local road authority that the backfilling of the excavation should be carried out in 1:12 weak concrete to prevent subsidence.

The excavation, extension of the drain from the saddle to the interceptor or first manhole within the site boundary, backfilling and making good to road and pavement is the responsibility of the building contractor but the work may be carried out by the local authority depending on local regulations.

7.00 Hot and cold water installations*

The distribution of cold water and the generation and distribution of hot water follows simple physical laws. These may be summarised as follows:

(i) Water finds its own level and is subject to gravity.

(ii) Water stored for distribution must be provided with a 'head', i.e. placed at such a level above the fitting served as to enable sufficient pressure to be generated to overcome the friction of the pipe and issue from it with sufficient force as to be acceptable to the user.

(iii) Warm water in a closed or open circuit will rise and displace cold water as warm water is less dense than cold and movement arises from the resulting pressure difference.

The proper working of these laws is subject to either sufficient storage being available at a suitable height above the topmost fittings or sufficient pressure being applied to water contained in the pipes.

Water supply services in a building conform to either one or the other of two standard systems:

7.01

(a) The Direct cold water supply system. In this the whole of the cold water is supplied direct to fittings from the service pipe. This system is in general use in districts where large high level reservoirs maintain a good main supply and pressure, e.g. Yorkshire and Lancashire. The only storage required is a small storage cistern to feed the hot water storage tank and in most cases this is provided by a combination cylinder in which cold water is stored in the top portion, automatically syphoned into the lower hot water cylinder when water is drawn off.

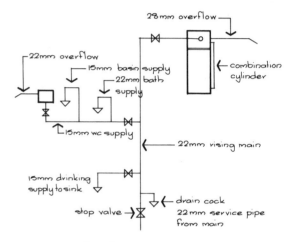

Direct system of Cold Water Supply

Advantage of this are:

(i) The combination cylinder may be positioned below the ceiling saving on pipework, simplifying installation and maintenance and providing additional safeguards against damage from frost.

(ii) Drinking water is available from all cold water outlets.

Disadvantages are:

(i) The risk of contamination of mains water by back syphonage due to either negative mains pressure or by outlets being submerged below the water level.

(ii) Lack of reserve should water supplied be cut off for repairs.

(iii) Lowering of supply pressure during periods of peak demand or drought.

7.02

(b) The Indirect cold water supply system. In this system, with the exception of a drinking water outlet at the sink, all fittings are served from a cold water storage cistern fitted with a cover and usually placed in the roof space to ensure sufficient 'head' or pressure to properly serve fittings on the first floor.

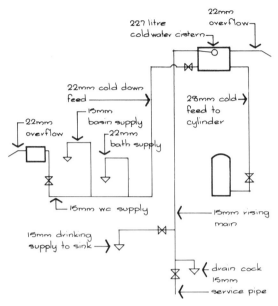

Indirect system of Cold Water Supply

The advantages of this system are

(i) it gives a reserve storage in case of mains interruption, and

(ii) reduces the risk of contamination from back syphonage.

In any event, the selection of the system to be used is determined by the water authority in the area.

7.03 Hot water supply

Hot water to domestic sanitary fittings is usually drawn from a hot water cylinder. The source of heat may be one of several alternatives as follows:

(i) By an electric immersion heater installed in the top of the cylinder.

(ii) By a gas circulator connected to the cylinder through short lengths of pipework.

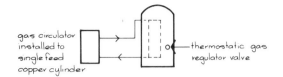

(iii) By means of a separate boiler fired either by solid fuel, gas or oil, sometimes incorporating separate circuits to utilise excess energy for the distribution of heat by radiators to various rooms in the house.

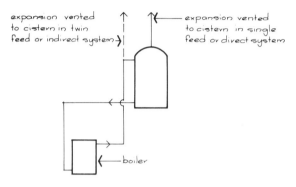

When hot water is drawn from the top of the cylinder it is immediately replaced with cold water from the storage cistern. It is essential to ensure that the supply pipe equals or exceeds the diameter of the draw-off to avoid loss of water in the cylinder.

7.04

(a) **The Direct hot water system.** In this system cold water from the cylinder flows through the heating

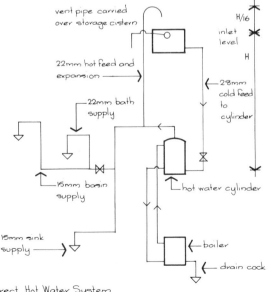

Direct Hot Water System

23

jacket of the boiler (or is directly heated by the immersion heater) where its temperature is raised forming natural convection currents which cause the hot water to rise and circulate. The hot water is replaced by colder water by gravity thus improving the circulation. Hot water is drawn off from the top of the cylinder from which a pipe is set to discharge over the top of the cold water storage to provide a relief vent.

This system is not suitable when radiators are to be included for heating or where the water in the area is hard. In the latter instance the pipes would become 'furred' up with lime deposits which occurs between temperatures of 50 and 70°C., the ideal temperatures for domestic hot water installations.

7.05
(b) Indirect hot water system. This system is designed to overcome the problem of 'furring' in which the cylinder contains a coil or annulus connected to the flow and return pipes from the boiler. The transfer of heat takes place within the cylinder and after the initial precipitation of lime within the primary circuit no further furring takes place as cold water is no longer being introduced into the boiler circuit.

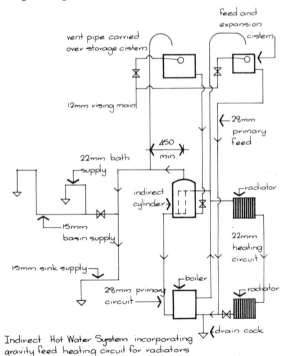

Indirect Hot Water System incorporating gravity feed heating circuit for radiators

The supply circuit from the cylinder follows the same pattern as for the direct system except that a separate feed and expansion tank is required to provide the initial supply for the boiler and primary circuit and to top up for evaporation. The size of the rising main may be increased to 22 mm in districts where the local water board consider the main pressure to be inadequate for the smaller size connection. This is also general practice in Scotland where the larger bore connection is always used.

7.06 Materials
Materials used for both cold and hot water installations are similar and can be considered together.

7.07 Pipework
Copper is generally used for internal pipework. Copper tubes are virtually inert to potable water and are strong enough to withstand deformation due to stresses caused by expansion and contraction. The tubes are easy to cut, manipulate and handle and are manufactured in metric dimensions to conform to BS 2871 Pt. 1. The dimensions in general use for domestic installations are:

12 (15 mm)	Main intake supply pipes, runs to W.C. cisterns, basins and sinks.
18 (22 mm)	Cold water down services to fittings runs to baths. Vent pipes and hot water draw off.
25 (28 mm)	Primary circuits and cold feeds to hot water circuits.

(Bore dimensions in parenthesis)

7.08 Control valves
It is essential to install controlling valves in cold and hot water installations to enable sections to be shut off for repair or extension. The valves are generally of brass with a wheel head coloured green for cold and red for hot water services. An arrow cast on the body of the valve indicates the designed direction of flow.

Valves are generally installed as a minimum in the following positions:

(i) On the cold down service next to the cistern.

(ii) On the cold down supply to the hot water service near the cistern.

(iii) On the hot down supply near to the bath draw off supply.

(iv) On the main service next to the cold water tap to the sink.

(v) On the main service next to the feed and expansion cistern.

If possible all valves should be placed where accessible either from the upper floor or immediately inside the roof hatch and should be labelled as to use. On no account should a valve be fitted to the draw off or expansion pipe.

7.09 Fittings

Copper tubing is manufactured with plain ends and needs separate fittings for jointing.

(i) Compression joints in which the square plain end of the tube is flared or belled by means of a special forming tool and compressed by the coupling nut against a loose thimble insert. A variation uses a swaging tool to roll a bead on the tube about 12 mm from the end which is compressed against the mouth of the fitting making a fluid tight joint. A third method uses a loose copper ring which grips the wall of the tube when the coupling nut is tightened. This latter fitting is only used above ground.

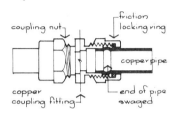

Typical Compression joint

(ii) Capillary joints are cold wrought from copper tube and incorporate solder rings which, when heated by a blow lamp, are drawn into and seal the narrow space between the outer wall of the tube and the interior of the fitting. A variation to this type of fitting is the end feed capillary fitting where separate solder wire is applied to the mouth of the fitting.

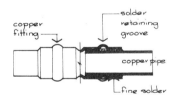

Typical capillary joint

7.10 Cold water storage cisterns

These are manufactured in a number of materials and a great variety of sizes to suit site requirements and conditions. Materials in general use are as follows:

(i) Galvanised mild steel to BS 417. These require additional internal protection from two coats of a black bituminous coating to BS 3416 to reduce the risk of chemical attack from chemicals found in the water supply in some areas.

(ii) Asbestos cement to BS 2777.

(iii) Polyolefin or olefin copolymer to BS 4213.

(iv) Polyester compound GRP.

Cisterns should be fully supported over their whole base area to prevent deflection when full of water. A boarded platform or wood wool slab with a layer of bitumen felt to separate the tank from the platform is ideal. Holes for services must be carefully cut, the swarf removed and the holes reamed smooth. Cisterns should be fitted with close fitting covers to protect the contents from dust and dirt.

The usual sizes for domestic installations are:

227 litres for cold water storage.
114 litres for feed and expansion storage.

7.11 Ball valves

Ball valves are required to control the supply of water from the main into the cold water storage cistern. In addition they are installed in W.C. cisterns to control the water supply to the fitting. Two types of brass ball valve are used:

(i) The *Piston valve* to BS 1212 Pt. 1 has a horizontal cylinder incorporating a piston with the float arm operating on a seating under the body of the valve.

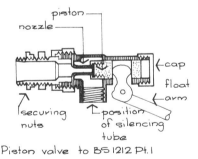

Piston valve to BS 1212 Pt. 1

(ii) The *Diaphragm ball valve* (Garston or BRS pattern) to BS 1212 Pt. 2 consisting of a simple diaphragm operated by the float lever arm on a tapered seating in the body of the valve.

Ball valves are made to suit varying pressures and should be selected to suit the local conditions.

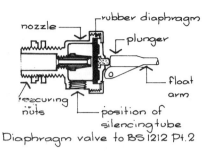
Diaphragm valve to BS 1212 Pt.2

Floats for ball valves are made from either copper or plastic and sized to suit the bore of the inlet pipe.

7.12 Overflows*
All cold water storage and W.C. cisterns must be fitted with an overflow or warning pipe set slightly above the normal water level. These should be one pipe size larger than the main inlet pipe and set to discharge into the open air.

7.13 Hot water cylinders
Hot water cylinders are made in both galvanised steel and copper although the latter are in general use today for domestic work. Copper cylinders are made in a number of sizes and capacities, as follows:

Copper direct cylinders to BS 699

116 litres	400 mm dia x 1050 mm high
144 litres	450 mm dia x 1050 mm high
166 litres	450 mm dia x 1200 mm high

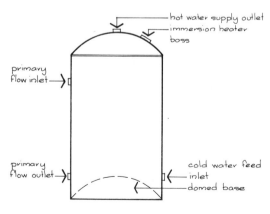
Copper direct cylinder to BS 699

Copper indirect cylinders to BS 1566 Pt. 1.

114 litres	400 mm dia x 1050 mm high
140 litres	450 mm dia x 2050 mm high
162 litres	450 mm dia x 1200 mm high

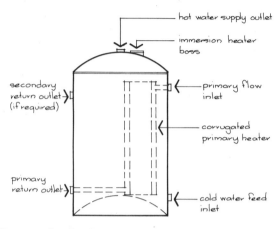
Copper indirect cylinder to BS 1566 Pt.1

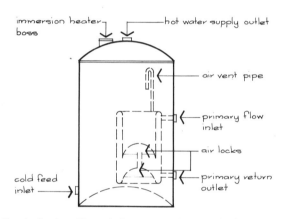
Single feed or 'Primatic' copper indirect cylinder to BS 1566 Pt.2 (see 7.02) for use with gas circulator etc. to replace separate feed required for boiler primary circuit

7.14 Boilers
Boilers for the production of hot water by means of indirect or direct cylinders are produced in a variety of sizes and thermal capacities to suit the sizes and layouts of particular installations. Energy in the form of heat is produced from a variety of fuels as follows:

(i) Solid fuel, usually either anthracite or one of a number of patent smokeless fuels.

(ii) Gas, either produced from coal or natural sources.

(iii) Domestic grade light fuel oil.

The boilers are generally simple and relatively inexpensive.

The combustion of solid fuel boilers is generally controlled by means of an automatic damper operated by a thermostat which closes the damper and restricts fuel consumption when the temperature in the boiler waterway reaches the optimum design temperature.

Gas-and oil-fired boilers operate on a similar system with a fuel cutout thermostat, but usually have an over-riding clock control which enables the user to restrict boiler operation to pre-selected periods of the day, thus saving fuel. Gas circulator hot water installations are generally designed to operate on the gravity flow system which does not need the provision of mechanical pump or accelerator to ensure proper movement of the heated water through the primary circuits.

7.15 Insulation

All pipes and fittings should be adequately insulated to ensure freedom from freezing. Where upper floors are insulated from roof voids all pipes in the roof space should be insulated with a minimum of 15 mm thick preformed fibreglass or similar pipe wrapping. Cold water storage tanks should be insulated with either,

(i) 50 mm loose vermiculite contained within a board or timber casing, or

(ii) 50 mm of strawboard or wood wool slabs, or

(iii) 25 mm expanded polystyrene.

If the tank can be located either over the linen cupboard or next to a flue in continuous use this will materially assist in preventing freezing in very cold weather.

In addition, all pipes concealed in voids in the structure or where exposed to the possibility of frost on external walls should be insulated. Cylinders and primary flow and return pipes show considerable fuel economy when properly and efficiently lagged.

8.00 The installation of sanitary fittings

Sanitary fittings are provided in all buildings to ensure personal individual and corporate hygiene. Types and patterns of fittings vary both with the use to which the building is put, minimum standards laid down by law in connection with certain activities and personal preference. Manufacturers provide their own particular patterns which conform to accepted standards and hydraulic principles in the various materials and finishes in common use.

8.01

Sanitary fittings fall into two main categories:

(i) Soil fitments which include W.C.'s, urinals, slop sinks etc, and

(ii) Ablution fitments which include baths, lavatory basins, bidets and sinks, etc.

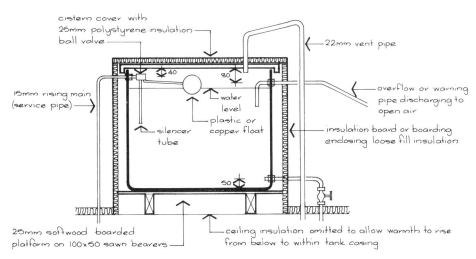

Typical cold water storage cistern with connections and insulation (pipe insulation omitted for clarity)

The materials used for these fitments are numerous and include:

Glazed fireclay	W.C.'s, urinals, slop sinks, lavatory basins, bidets and sinks.
Vitreous china	W.C.'s, lavatory basins, bidets.
Porcelain enamel cast iron	Baths.
Porcelain enamel steel	Baths, vanitory basins.
Coloured plastics	Baths, vanitory basins.
Stainless steel	Urinals and sinks.

The main qualities required from all these materials when used for sanitary fittings is that they shall meet the following criteria:

(i) Finish shall be hard, smooth and impervious.

(ii) The finish shall be free from corrosive tendencies.

(iii) The design shall be self-cleansing, clean and simple.

(iv) Working parts shall be reduced to the minimum to ensure effective emptying in a simple quick action.

8.02

Sanitary fittings provided in small domestic works are as follows:

(a) *The W.C.* which is manufactured in two patterns:

(i) Washdown pattern which flushes by means of the gravity overflow discharge principle from water contained in a flushing cistern at a higher level than the pan.

Typical washdown wc suite

(ii) Syphonic pattern from which the contents are removed by syphonic action induced from water discharged at will from a flushing cistern also at a higher level than the pan.

Typical syphonic close coupled wc suite

The W.C. must be provided with water through a 15 mm supply pipe from the water storage cistern which is connected to the screwed tail of the ball valve fitted inside the flushing cistern. It is usual to provide an isolating stopcock on this supply. In addition a 20 mm overflow pipe must be fitted to the flushing cistern to discharge to open air as a safety device against overfilling.

The flushing cistern is either manufactured integral with the pan, when it is known as a 'close couple suite', or connection is made between the two by means of a flush pipe. This is connected to the outgo of the flushing cistern by a coupling and to the pan by means of a cone connector.

Connection between the pan and the branch pipe socket of the one pipe system vertical soil and vent shaft (10.19) is by means of a self-sealing plastic pan connector.

8.03

(b) *The Bath*, manufactured in two main sizes:

1700 x 700 x 545 mm high (BS 1189:1972)
1800 x 800 x 550 mm high

The height of a bath is variable either on account of adjustable feet (cast iron and steel), or the height of the fixing and supporting cradle (plastic). The free sides or ends of the bath are enclosed by panels formed from a variety of materials, most of which need some form of rough framing in timber to support or secure.

Cold and hot water supplies to the taps are 20 mm and are connected direct to the taps with an appropriate connector. Baths are usually supplied complete with a chromium plated outlet, plug and chain and flexible overflow and these must be fixed in position and interconnected to ensure safe discharge of water in all circumstances. Incorporated in the waste and overflow is a 38 mm. diameter bath trap with a 75 mm seal suitable for connection to one pipe system soil and vent shaft (10.19).

8.04
(c) *Lavatory basins*, manufactured in a great variety of shapes and sizes of which the following are typical:

   700 x 500 mm wide (BS 1188:1974)
   600 x 500 mm wide (BS 1188:1974)

Basins are either fixed to steel supporting brackets plugged and screwed to the wall or are provided with integral lugs which are grouted into recesses formed for them in the structure of the building, or are supported on complimentary pedestals.

Cold and hot water supplies to the taps are 15 mm and are connected direct to the taps with an appropriate connector. The overflow to the basin is formed integrally in the body of the unit connecting to the waste outlet in the bottom of the basin. The chromium plated outlet incorporates a slot in the side of the outlet pipe to collect water from the overflow slot. A 32 mm diameter 'P' trap with a 76 mm seal suitable for connection to the one pipe stack is fitted to the basin waste outlet. As this is usually visible even when the basin is provided with a pedestal the waste is usually of chromium plated brass. Traps can be made of either plastic or c.p. brass.

8.05
(d) *The Sink* is usually either a stainless steel pressing or a pressing of vitreous enamelled sheet steel. The sizes of sink and combined drainer are manufactured to suit standard kitchen unit fittings comprising drawers and cupboards, and as such may be of some size and variety. For general work the sink tops have either a single or a double drainer and are usually sized as follows:

Single drainer sink top	1070 x 530 mm wide
Double drainer sink top	1600 x 530 mm wide (BS 1244: Pt. 2 1972)

Main drinking and hot water supplies to the taps or approved pattern non-contaminating mixer taps are 15 mm and are connected direct with the appropriate connector. As with the lavatory basin the overflow and the waste outlet are interconnected and are a supplied fixture. A 50 mm seal trap is required unless the sink is to be connected to a one pipe soil and vent stack, in which case a 76 mm deep seal 'P' trap will be required.

9.00 Heating domestic structures
The principal requirements of a heat source is to

(i) provide the maximum heat possible from the quantity of fuel burnt for the benefit of the building occupants;

(ii) ensure that combustion takes place without any risk of fire spreading to the surrounding structure;

(iii) ensure that smoke and products of combustion are removed and dispersed without danger to health or annoyance to the building occupants and surrounding properties.

9.01
Combustion requires air and the replacement of air consumed is a necessary requirement for an efficient fire. Assuming that the flue is clean there are three ways in which poor combustion can occur and the efficiency of the fire be impaired:

(i) Insufficient air replacing that consumed by initial combustion.

(ii) Resistance to the flow of hot gases and smoke through the flue caused by poor flue design.

(iii) Downdraught, resulting from the build-up of pressure at the top of the flue due to excessive height of neighbouring buildings, trees, etc.

9.02
The Building Regulations Parts L and M* apply stringent controls to the installation of fixed heating appliances and the design of their structural requirements and flues for domestic buildings. Firstly the regulations classify the type of appliance and its rating in kW. So far as gas fuel is concerned, the rating is controlled by the heat content of the gas before combustion. With solid fuel and oil the rating is concerned with heat output.

		Rating in kW	
Class	Fuel	Input	Output
1	Solid fuel or oil		Under 45 kW
2	Gas	Under 45	

The application of the regulations to particular classes is described later.

9.03*

The general requirements of the Building Regulations in respect of flues, hearths and fireplaces as applicable here are as follows:

(i) Flues must not allow products of combustion to escape into the building (L2(2)).

(ii) The outlet of any flue must be so positioned and designed so as to prevent any discharge from entering an opening in the building (L2(4) and L13).

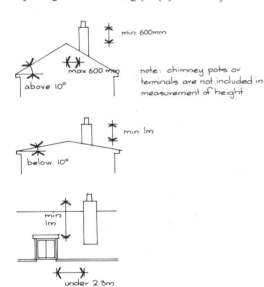

(iii) Flues serving solid fuel or oil burning appliances (Class 1) must be provided with a cleaning opening fitted with a close fitting incombustable cover unless the flue can be cleaned through the appliance itself (L2(6)).

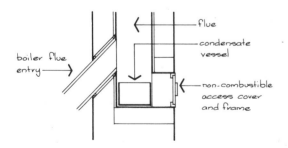

(iv) Flue linings to Class 1 appliances are to be rebated and socketted linings made from either

(a) kiln burn aggregate and high alumina cement, or

(b) rebated and socketted linings of clay to BS 1181: 1971, or

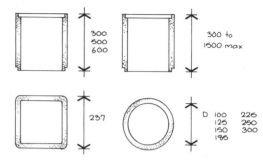

Flue linings to BS 1181

(c) glazed clay pipes and fittings to BS 65 and 540 Pt. 1: 1971, socketted, imperforate and acid resistant (L6).

(v) Flue linings to Class 2 appliances may be one of a number of deemed to satisfy constructions, but generally installed are

(a) sheet steel flue pipes and fittings to BS 715:1970, coated on the inside with acid resistant vitreous enamel, or
(b) asbestos cement pipes and fittings to BS 569: 1968, coated on the inside with an acid resistant compound, or
(c) a flexible steel flue internally coated as (a) installed in a brick flue size 225 x 225 mm with the void around the steel flue filled in with an inert insulant. (This method is in general use for gas boiler installations in existing structures).

9.04

The construction of a fireplace*, recessed for Class 1 appliances or open domestic fires, must be carried out in bricks, blocks or *in situ* concrete. The thickness of the back wall must be carried up the full height of the recess without taking into account the thickness of any fire brick which may be provided.

The construction of the hearth must be wholly of incombustable material and the hearth must be constructed to minimum dimensions.

Provision must be made for an adequate supply of air for combustion and the construction of a pit or sump under the appliance must be carried out wholly in non-combustible material.

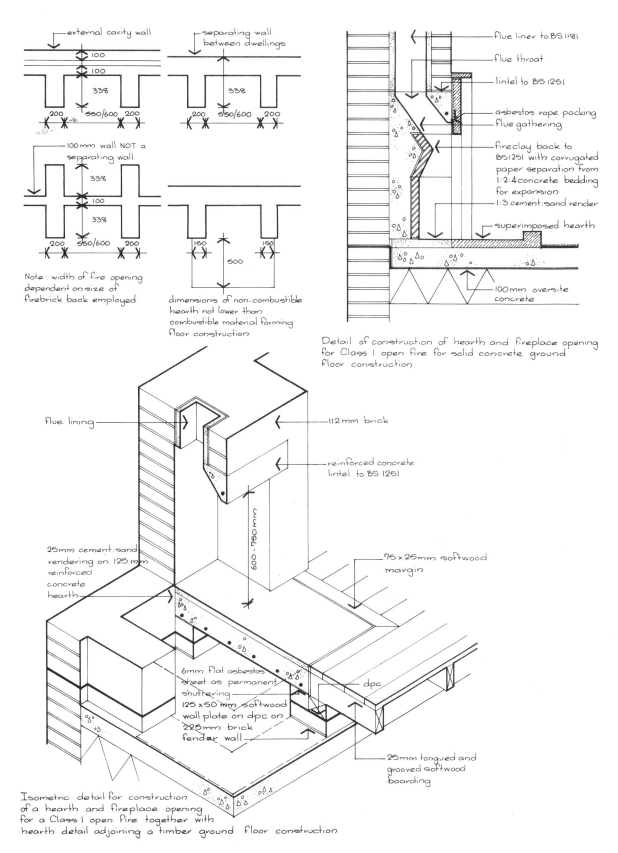

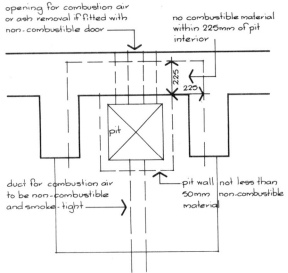

Plan of pit and statutory requirements relating to construction

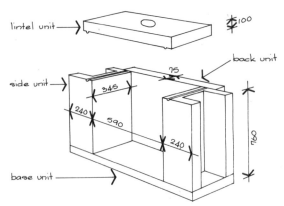

Precast unit for room heaters manufactured from Leca expanded clay aggregate

(iii) Free standing chimney unit constructed of precast units of cellular construction in cement bonded refractory material especially suitable for internal flues.

9.05

A number of precast units are now available which not only comply with the regulations but also provide fast and economic installation of both hearth and flue in both new and existing buildings. They are available in three main categories:

(i) Units for open fires incorporating a combined lintol and throat unit for connection to the flue.

(ii) Units for room heaters which incorporate a perforated lintol unit for connection to the flue.

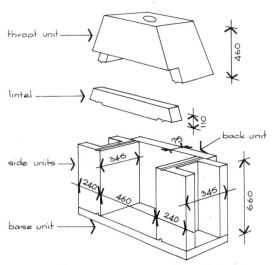

Precast unit for open fires manufactured from Leca expanded clay aggregate

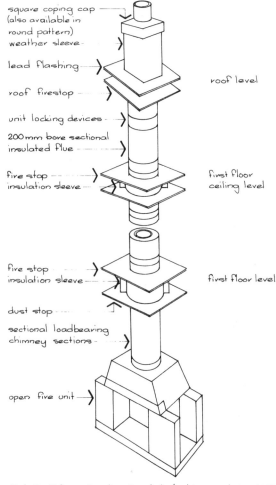

Detail of free-standing insulated chimney to BS 4543 for class I appliances, etc.

Units (i) and (ii) are constructed of light weight clay aggregate panels and are designed to be used in conjunction with an insulated prefrabricated chimney to BS 4543: 1970, *Factory-made insulated-metal chimneys* (L22)*.

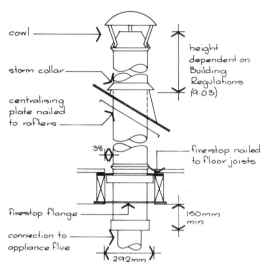

Detail of prefabricated insulated metal flue to BS 4543

9.06

The passage of flues through floors, ceilings and roofs often requires that the construction be modified somewhat to ensure continuity*. With timber construction this usually takes the form of 'trimming' the timbers around the opening to carry intermediate members.

With floors this is usually a simple procedure involving the use of a short trimmer joist secured by means of timber connectors.

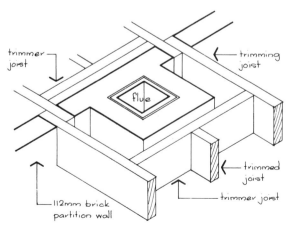

Detail of trimming at first floor timber joists around single flue stack (see 4.04 for jointing details)

A similar method is used for ceilings except that as the load on the members is negligible, the joint is usually a butt joint secured by splay cut nailing.

The trimming of roof timbers around the chimney stack varies due to the positioning of the chimney on the roof slope. When the chimney penetrates the ridge, the ridge board is cut close to the stack and the opening trimmed between the closest pair of rafters on either side.

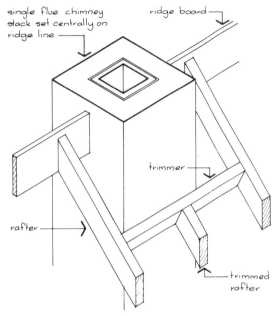

When the chimney stack penetrates the roof slope the opening must not only be trimmed but the back of the stack must be provided with a back gutter (or lier board) which, when properly weathered with lead, ensures the prevention of water penetration from stormwater running down the back slope of the roof.

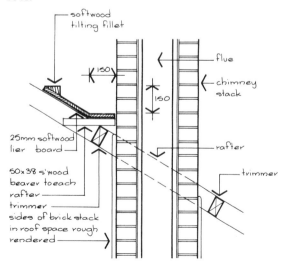

9.07

The opening formed in the roof weathering by the stack must have its edges sealed from water running down the stack, running down the roof slope and penetrating the material from which the stack is constructed. This protection is provided as follows:

(i) By inserting a Code 4 lead 'safe' or d.p.c. through the whole thickness of the stack, perforated for the flue and incorporating a front apron flashing, dressed down over the roof weathering to seal the front edge gap. (Note that the d.p.c. is turned up around the flue perforation and internally in the roof space to prevent water from above 'spilling over' into the structure below).

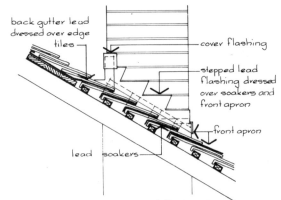

Code 4 lead soakers and stepped flashing to side of stack

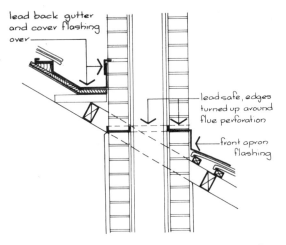

Code 4 lead safe through single flue stack perforated for flue and incorporating front apron flashing dressed down over roof weathering, and back gutter dressed over tilting fillet and at either side over roof covering

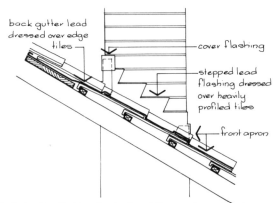

Code 4 side flashing to stack omitting soakers, flashing dressed over heavily profiled roofing tiles eg pantiles

(ii) By the provision of a Code 4 lead back gutter (9.06) dressed up under roof weathering to effect a seal at the upper edge and over at the side where the lead is dressed tightly to the roof surface.

(iii) By the provision of small Code 4 angled lead soakers, fixed one under each tile or slate down the side slope of the stack, the open top edge secured and sealed by a similar lead stepped flashing. (Note that when heavily profile tiles such as pantiles or interlocking tiles are used on the roof the stepped flashing incorporates an over flashing which is dressed over the weathering roofs of the tiles, the soaker being consequently omitted).

9.08

The top of the chimney must be protected against excessive rain and exposure. In brick stacks, the mortar should be 1:3 cement and sand and, if possible, all brickwork should be 225 mm in thickness; half brick stacks rarely being suitable and quickly deteriorating. The top of the flue should be sloped off or weathered and the top courses set to either oversail to throw water clear of the brickwork below or a concrete capping provided.

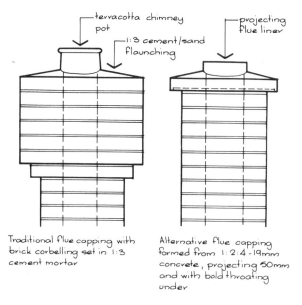

Traditional flue capping with brick corbelling set in 1:3 cement mortar

Alternative flue capping formed from 1:2:4 - 19mm concrete, projecting 50mm and with bold throating under

The flue liner should project 50 mm above the weathered top of the stack to prevent water running down the interior of the flue.

9.09*

Gas appliances (Class 2) of the small boiler or room heater type usually have their flues terminated by one of two methods, either

(i) By means of an asbestos cement flue ridge terminal connected to a preformed flue pipe so designed to allow free discharge, minimise down draught and prevent the entry of any matter which might restrict the bore of the flue, or

(ii) by means of an outlet set in a chimney connected to precast concrete flue blocks 100 mm thick which can be incorporated in brick or block walls to which proper bonding can be provided. These flues and terminals are most suitable for gas fires and similar single appliances.

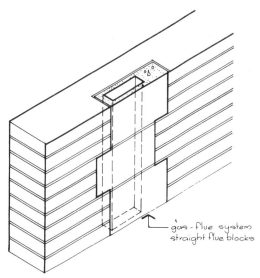

system incorporates raking blocks, offsets, and a range of terminal fittings

Detail of flue blocks set in 225mm brick wall

In addition larger boilers are often mounted against external walls, drawing air for combustion and discharging products of combustion through a 'balanced flue terminal' fitted to a flue passing direct from the back of the boiler through the wall to the open air.

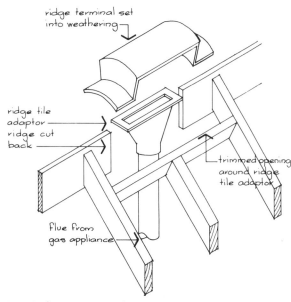

Detail of ridge terminal to flue serving Class 2 heating appliance

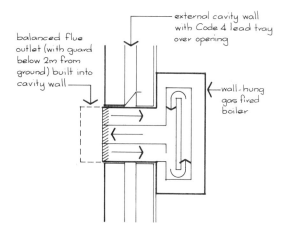

10.00 Domestic foul and stormwater drains

Drainage is a system of pipework to convey the discharge from sanitary fittings to a suitable and satisfactory disposal installation. In most urban areas disposal is by means of sewers laid and maintained by the local authority discharging into sewage treatment plants. Alternatives are small self-contained treatment plants or collection tanks conforming to CP 302. The system should be self-cleansing and operate with the minimum of maintenance.

In most installations drainage takes two forms:

(i) The system of pipework fixed above ground to collect the discharge from the fittings and in addition to provide proper ventilation facilities to the whole drainage system.

(ii) The system of pipework and fittings provided both at ground level and below the surface to collect the discharge and dispose of it to the sewer.

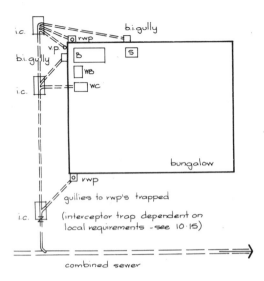

Typical combined drainage system

10.01

The layout of small domestic drainage schemes depends on a number of factors each varying from one scheme to the next.

(i) The number of discharge points.

(ii) The positions of the discharge points in relation to the point of disposal (the sewer).

(iii) The type of drainage system installed in the road by the local authority.

There are three drainage systems which are in general use in this country depending on a number of factors which are not relevant in this book. The local authority will determine the method used for the basic layout for each individual scheme dependent on their sewer system. These systems are as follows:

(i) *The combined system* in which all drains whether serving sanitary fittings or stormwater disposal discharge into a common sewer. This system is easy to maintain and overall flushing is regular. Its disadvantage is the quantity of water to pass through the treatment plant in wet weather.

(ii) *The separate system* which is the most common employed. One sewer receives the surface water and conveys this to a natural watercourse where it is discharged without treatment. The second sewer receives the discharge from the sanitary fittings and passes this through the treatment plant.

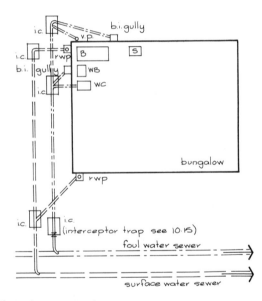

Typical separate drainage system

(iii) *The partial separate system* in which two sewers are used, one carrying surface water only and the other acting as a combined sewer. The amount of surface water induced into the latter can be regulated by design to ensure regular flushing without overloading the treatment plant.

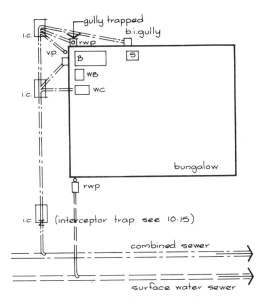

Typical partial separate drainage system

10.02
The Building Regulations (Part N)* lay down the basic principles of good drainage to be followed in all systems:

(i) Materials used for drains must have adequate strength and durability (N4 and 1Q)*.

(ii) The minimum diameter of the drains shall be

(a) 100 mm for soil drains.
(b) 75 mm for stormwater drains (N10)*.

(iii) All parts of the drain must be accessible for inspection and cleansing (N12)*.

(iv) Drains should be laid in straight runs between access points (N12)*.

(v) Drains must be laid to a gradient or fall which will render them self cleansing (N10)*. The fall is related to the rate of flow, velocity of contents and diameter of drain. Small domestic drains are intermittent in action and their flow is thus irregular. The minimum gradient may be calculated by the following formula:

Gradient = diameter of pipe (mm)/2.5 (Maguire's rule)*

This shows that the minimum gradient for a 100 mm pipe is 1 in 40 with a velocity of 1.4 m/sec. (See also CP 301: 1971, Section 3.5)

(vi) All drain inlets must be provided with a trap with a minimum seal of 50 mm. Traps are usually provided with the sanitary fitting (8.03 *et seq*) except where discharge is to gullies. R.w.p.'s need not be provided with a trap unless they are connected to a soil drain or sewer (N13)*.

(vii) Inspection chambers (manholes) are required at changes of drain direction and gradient where these prevent the drain from being readily cleansed (N12)*.

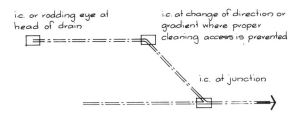

Manholes must also be placed within 12.5 m of drain junctions and not more than 90 m apart (N12)*.

(viii) Junctions between drains must ensure that the incoming drain joins at an oblique angle to the direction of flow (N16)*.

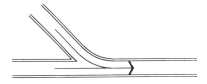

(ix) Drains under buildings must be either

(a) encased in 150 mm of mass concrete, *or*
(b) be laid using cast iron pipes.

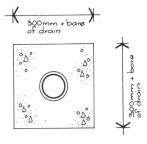

In any event flexible connections must be provided to the drain on either side of the building to allow the building to settle under load without fracturing or disturbing the drain (N15)*.

(x) Drain trenches occuring within 1 m of the foundations of the building and below the foundation level must be backfilled with concrete up to the level of the underside of the foundations. Drains more than

1 m from the foundations are backfilled with concrete to a depth equal to the distance of the trench from the foundation less 150 mm (N14)*.

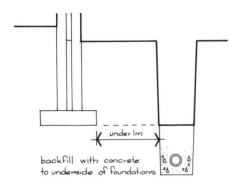

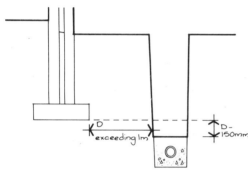

(xi) The minimum invert level (depth below ground of the bottom of the pipe bore) of a drain should not be less than 450 mm to avoid damage by ground movement and traffic.

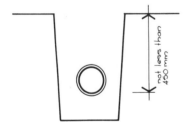

10.03 Drain pipes

Drain pipes are of two principal types, rigid and flexible. Materials conforming to these are as follows:

(i) Rigid:
  clay or stoneware
  cast iron

(ii) Flexible:
  pitch fibre
  UPVC

Whereas at one time it was considered that rigidity in construction was a prerequisite of sound drainage construction, this policy is now no longer considered satisfactory due to settlement which causes the jointing of rigid drain runs to fail. Present practice utilises either rigid or flexible pipes with flexible joints which allow settlement to be taken up without loss of contents at disturbed joints.

10.04

(a) **Vitrified clayware pipes** (to BS 65 and 540 Pt. 1: 1971). The BS does not insist on these pipes being glazed but most are finished in this way, the glaze being either salt or ceramic. Two patterns of clayware pipes are in general use for both soil and stormwater drains, British Standard pipes used for both systems and British Standard surface water pipes for stormwater only. These latter can be obtained perforated for sub-soil drainage (23.05). The patterns available are as follows:

Spigot and socketted pipes and fittings produced in a variety of sizes and lengths to suit requirements, e.g.

Nominal bore (mm)	*Effective length excluding socket (mm)*		
75	300	600	
100	300	600	900
150			
225	1000	1200	1500

Jointing can be effected either by rigid or flexible methods. The traditional rigid joint comprises tarred gaskin caulked into the joint which is then fully filled with cement and sand mortar 1:2 with a bold splayed collar to seal the joint.

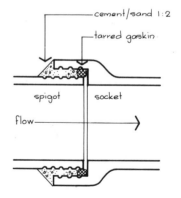

Flexible joints use sealing rings or 0 rings recommended by the manufacturers of the pipes. In any event these flexible joints must be used close to the wall face when pipes pass through walls or are connected to manholes (10.02 (ix)).

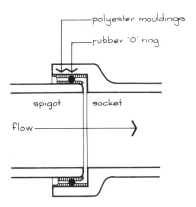

typical flexible joint using clay drain pipes

Pipes are now produced with square ends which are connected with a push fit coupling moulded in polypropylene incorporating rubber sealing gasket. Adaptors are made to enable connection to be made to standard terminal fittings. A proprietory lubricant is used to make a tight fitting flexible joint.

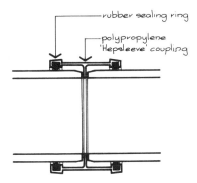

flexible coupling joint to square end clay pipes

10.05
(b) Cast iron pipes (to BS 437 Pt. 1: 1970). Cast iron pipes are suitable both for soil or stormwater use and are made in the following sizes and lengths to suit requirements.

Nominal bore (mm)	Effective length excluding socket (mm)		
75			
100	1830	2740	3660
150			
225			

The pipes are coated with a mixture formed with a bitumen base to provide protection in the earth.

Fittings comprise bolted and airtight covers in a variety of standard patterns to suit most branch jointing situations fitted with sealed and bolted airtight covers for use within buildings. Jointing is made by means of tarred gaskin or lead wool well caulked in with a run lead joint or the use of a cementitious compound well caulked into the joint.

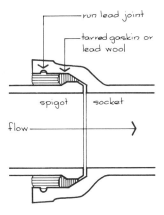

rigid joint to BS 437 cast iron pipe

10.06
(c) Pitch fibre pipes (to BS 2760 Pt. 1: 1966). Pipes manufactured from pitch fibre are made in a variety of diameters and two standard lengths as follows.

Nominal bore (mm)	Effective lengths (mm)	
75		
100		
150	2400	3400
200		
225		

Two distinct jointing techniques are used:

(i) The butt joint in which pipes are joined by connecting two plain end pipes with socketted polypropylene couplings and two rubber D rings (flexible).

(ii) Tapered pipes are connected to a similar pipe by means of a pitch fibre or polypropylene tapered coupling, usually by driving one pipe onto another by a wooden dolly (rigid).

Adaptors are provided to joint pitch fibre pipes to vitrified clay fittings.

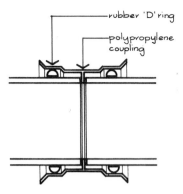

flexible joint to BS 2760 pitch fibre pipes

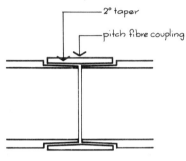

rigid joint to pitch fibre pipes

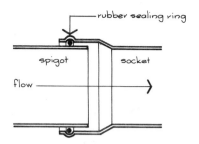

flexible ring seal joint in uPVC pipes

10.08 Drain laying

Drains are laid underground in trenches excavated with their sides supported in a similar way to foundations. The only variation is that the bottoms are graded to fall to suit the design of the system. Drain trenches should be opened for as short a period as possible and preferably the minor runs should be installed as early as possible in the works to avoid disruption caused by open excavations.

Drainlaying and the bedding of drains depends on whether the pipe and joint is rigid or flexible. The Building Regulations (N10)* require that the drains be of sufficient strength and durability and be so jointed as to remain watertight under all conditions including that of ground movement.

Pipes with socketted joints are laid from the bottom of the drain run with the socket against the flow, each pipe being laid to the correct fall, aligned in straight runs and the bore centralised. The tarred gaskin forms the aligning medium in socketted pipes; in flexible joints it is the coupling which ensures alignment and pipes jointed in this way can be laid in either direction.

10.09

Where the subsoil is firm and stable the traditional rigid concrete bed and haunching is used with rigid jointed pipes of vitrified clay or cast iron. The advantage of this method is that the excavated spoil can be returned as backfilling, often showing economy in costing.

10.07

(d) Unplasticised PVC pipes (to BS 4660: 1973). uPVC pipes and fittings are produced, like pitch fibre pipes, in one grade only both for soil and stormwater drains. uPVC pipes are produced in a variety of sizes and lengths, those suitable for small works being as follows:

Nominal bore (mm)	Effective length excluding socket (mm)	
	Ring seal socket	Solvent socket
110	1000 3000 6000	1000 3000
160	— 3000 6000	— 3000

Jointing of pipes is usually effected by means of a ring seal which allows movement to take place without joint failure.

Pipes are generally jointed on the surface and lowered into the trench. Certain joints between pipes and fittings have to be made by the solvent cement method which incorporates a spigot and socket joint. Connectors are available to allow connection between uPVC pipes and cast iron and vitrified clay pipes and fittings.

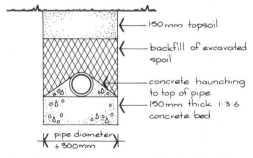

traditional bedding for rigid pipes with rigid joints in firm, stable subsoil

10.10

Where rigid pipes with flexible joints are employed movement joints should be used, the concrete bed being provided with construction joints at 5 m centres to allow movement in the subsoil.

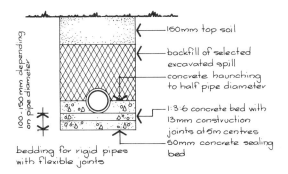

bedding for rigid pipes with flexible joints

10.11

Flexible pipes with flexible joints can be laid with a wholly flexible bed incorporating a granular material which is hand tamped under and around the pipe. The material used generally has a particle size of from 5–20 mm and often has to be imported to site for the purpose.

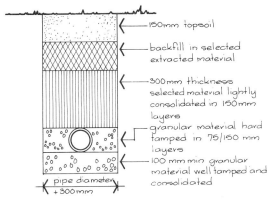

granular bedding for flexible *or* rigid pipes with flexible joints

10.12 Inspection chambers

These chambers are often called 'manholes' and are constructed as required by the Building Regulations (N12)* to provide access and inspection facilities to every length of drain as follows:

(i) At change of direction.

(ii) At change of gradient.

(iii) At junctions (unless other methods of access for cleansing are provided).

(iv) At the head of each drain unless a clearing eye is provided.

The size of manholes depends on the angle and depth of the main drain run and the number and positioning of branches. A guide for manhole sizes is as follows:

Invert depth (mm)	Size L x B (mm)
Up to 900	750 x 700
900 to 2700	1200 x 750

These dimensions are for access only, the length of the manhole should be calculated as follows where branches are provided:

Size of branch (mm)	Allowance for connection (mm)
100	300
150	375

(plus 300 mm to each for downstream end of angle of entry).

The width of manholes should allow 600 mm for benching plus the bore of the pipe (mm).

Manholes* constructed in clay engineering bricks to BS 3921 Pt. 2. Class B in cement mortar 1:3 using sulphate resisting cement where ground conditions require should have walls 225 mm thick fair pointed internally. The wall thickness can be reduced to 112 mm for the top 900 mm. The manhole should be constructed on a 1:3:6 concrete base 150 mm thick and provided with a similar cover rebated for a cast iron cover and frame.

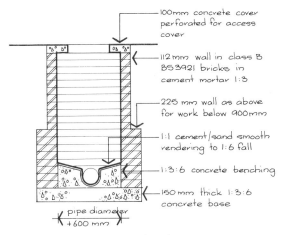

typical brick inspection chamber

Precast concrete sectional manholes to BS 556 are useful in ground with a high water table. They should be surrounded in 150 mm concrete 1:3:6 and set on a similar concrete base. Channel and connections are manufactured to order by the supplier.

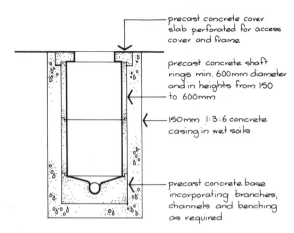

Precast concrete inspection chamber to BS 556

Cast-iron bolted and sealed junctions for use with cast iron drains under buildings are generally set on a bed of concrete with enclosing walls of brick; the cover and frame provided being of a type suitable to allow the floor finish to be laid in it and improve appearances.

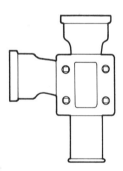

Cast iron access junction with door screwed with gunmetal bolts and nuts and greased felt washers; other patterns available for particular situations

Glass fibre reinforced plastic (GRP) and uPVC manholes can also be obtained and are fixed in accordance with the manufacturer's instructions and connected to uPVC drain runs as previously described.

10.13*

Channels in brick manholes are formed from half round vitrified clay with tapers and branch channels as required of three quarters section and set to discharge with the flow of the main drain. The side benching should be 300 mm wide in fine concrete and rise vertically off the top edge of the channel to at least the soffit of the outlet and then sloped back to the manhole wall at an angle of 1:6. The benching should be trowelled smooth.

10.14

Access covers and frames should be of cast iron, generally light patterns to BS 297 being suitable for domestic work, single seal types being used. The covers should be bedded in cement and sand 1:3 onto the concrete cover of the manhole and the covers bedded in the frame in non-hardening grease.

10.15*

Some local authorities require that private soil drains shall be isolated from the main sewer system by a trap incorporating a water seal. This trap is known as an interceptor trap, provided of the same nominal bore of the drain run and incorporating a clearing branch sealed with a stopper to allow clearing of the drain between the manhole and the sewer connection.

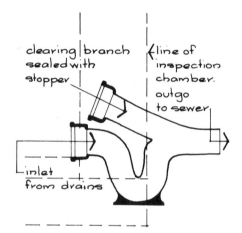

Section through typical interceptor installed in wall of manhole

As interceptor manholes are sometimes of some depth it may be necessary to build in cast iron step irons in the wall to facilitate access.

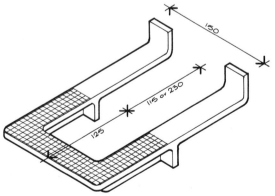

Malleable cast iron manhole step iron to BS 1247

10.16 Ventilation of drains*

All drains must be ventilated at their head, usually from the top manhole by means of a ventilation pipe which is carried above any opening into the building so as to prevent the discharge of any foul air in a position where it can gain access. In addition the top must be protected by a durable wire cage or 'balloon'. Generally the vent pipe is incorporated in the soil stack which is described later (10.17).

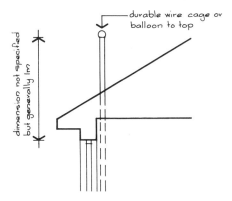

The minimum diameter permitted is 76 mm and the pipe is constructed either of cast iron, asbestos cement or uPVC.

To ensure a through draught from the interceptor (or bottom) manhole a fresh air inlet (f.a.i.) is provided and jointed into the manhole wall at high level. The inlet incorporates a flap ventilator which prevents the egress of foul air while permitting fresh air to enter the system. A f.a.i. is essential if an interceptor trap is provided – omitted if no interceptor is installed.

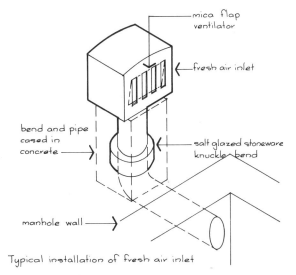

Typical installation of fresh air inlet

10.17 Soil and ventilating stacks

The choice of materials and the design of one pipe or 'single stack system' soil and ventilating stacks for small dwellings is covered by CP 304. Materials in use include:

(i) Cast iron, usually described as LCC pattern and coated with a black bituminous preservative known as Dr. Angus Smith's solution, and manufactured to BS 416:1967. Jointing is similar to cast iron drains (10.05).

(ii) Asbestos cement pipes, sometimes coated with a black bituminous composition and jointed with a cementitious fibrous jointing compound.

(iii) uPVC jointed either by solvent cementing (incorporating expansion joints) or by synthetic rubber 0 ring joints, manufactured to BS 4514.

Whatever the material used, the joints must be water and air tight and remain so in use. No jointing material must intrude into the bore.

10.18

The sizes of pipes used for small installations are as follows:

Purpose	Diam. of stack (mm)
2 storey dwelling with max. branch dia. of 76 mm.	76 for collection and discharge
2 storey housing	89 of water only
Flats up to 5 storeys	100

10.19

The design of one pipe or single stack soil pipes should be in accordance with the following principles:

(i) A bend at foot of stack can cause back pressure at the lowest branch. Care must be taken to keep root radius of bend in excess of 150 mm in addition to the minimum invert dimensions shown. Large radius bends or two 135° bends can also be used.

(ii) W.C. branch connection can cause induced syphonage lower in the stack when discharged. Connections should be swept in the direction of flow and maximum branch lengths of 1500 mm employed.

(iii) Basin wastes can induce self syphonage. Falls to wastes should not be less than 1/48.

(iv) Bath wastes should be connected to a one-pipe stack either

(a) above level of centre line to W.C. branch, or
(b) below a level 200 mm below this.

(v) Sink wastes should have a fall of not more than 1/24 to avoid blockage.

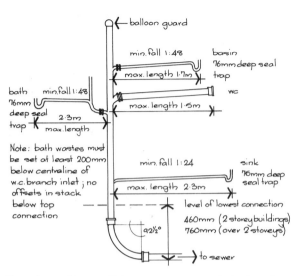

Typical installation of domestic one pipe soil stack and branch connections

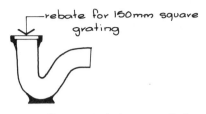

Yard gulley with 'P' trap outlet

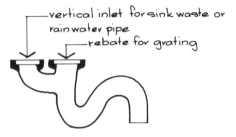

Back inlet gully with 'S' trap outlet

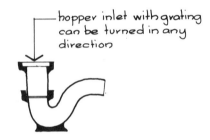

Reversible gully with square hopper inlet and 'P' trap outlet

Traps to fittings should be 76 mm deep seal P-type traps as described before in the installation of sanitary fittings (8.03 et seq).

One-pipe stacks are generally fixed internally to reduce the risk of frost damage and are provided with a wire or plastic dome at the top to prevent blockage. (See also BRS Digest 80).

10.20 Stormwater drains

The collection of stormwater from roofed areas has been discussed previously in *Building Technology 1* (22.25). The rainwater drainage system is to collect stormwater from the bottom of the rainwater pipes and paved areas and dispose of the water either to public stormwater sewers or back into the ground.

The method of transfer to sewers follows, in principle, methods used for soil except that the drains are not vented or isolated from the sewers by interceptor traps. Gullies provided in paved areas and at the feet of r.w.p.'s are, however, provided with a trap to ensure a water seal and the r.w.p.'s are connected into back inlets provided to discharge the water below the gulley grating.

In addition, branches do not need to be provided with manholes for inspection or rodding purposes, these are only generally provided at changes of direction.

10.21

In areas where the subsoil is suitable, e.g. in chalk or similar well drained material, advantage is taken of the high natural absorbtion of the subsoil to discharge stormwater directly back into the ground. This is effected by connecting the r.w.p. direct to the drain socket bend at ground level and running the drain a minimum of 5 m clear of the building where it is connected to a soakaway. This may be constructed in one or other of two ways depending on the situation of the property. For rural areas a hole is excavated about 1 m below the invert of the inlet pipe and filled with broken brick hardcore to a depth 300 mm over the top of the pipe and sealed from overlying soil by a 75 mm layer of concrete. The inlet pipe is turned down into the hardcore to prevent blockage. The size of the soakaway is determined by the volume of flow and the porosity of the subsoil.

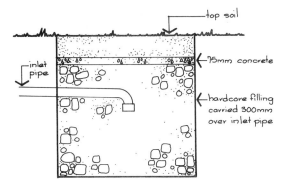

Typical soakaway in well-drained subsoil

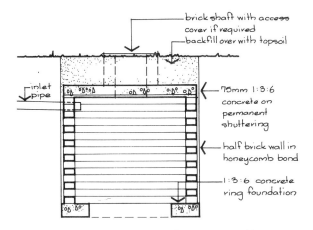

Typical brick steined soakaway for domestic buildings

In urban areas the soakaway is usually constructed by forming a ring of brickwork in honeycomb bond onto a 150 mm concrete ring foundation. The inlet pipe discharges through the wall and the soakaway is provided with a cover of concrete cast onto a corrugated asbestos sheet permanent shutter.

Sometimes an access cover is provided of precast concrete fitted with a lifting hook or a cast iron manhole cover and frame. Usually the cover is set down below ground level and topsoil is filled in over the top.

10.22 Testing drains

The Building Regulation (N11)* requires that drains and private sewers must be capable of withstanding a suitable test for water tightness after completion. The test is carried out by officers of the local authority. Testing is carried out in accordance with CP 301 : 1971, and provides for both water and air testing. In general the water test is applied to underground drainage and comprises filling the system in sections with water and inspecting for loss over a prescribed period of time.

Air testing is used for soil and ventilating stacks, both ends of the stack being plugged off and air pumped in from a hand pump until a prescribed pressure of 50 mm head of water is indicated on a U tube gauge connected to the system, all W.C. and sanitary fitting traps being filled with water. A fall of 12.5 mm in the pressure over a period of 5 minutes indicates a faulty stack.

Any defective section of drain or faulty joints in the stack must be satisfactorily repaired and the whole retested.

Module C Finishes and Finishings

11.00 Finishes and finishings
The term 'finishes' applies to a number of materials in which either

(i) they have a smooth surface suitable to receive a decorative application or finishing, or

(ii) the material is itself provided with its own integral decorative finish.

Applied finishes, to which (i) refers, are generally applied over carcassing or superstructure works to provide a better finish than originally provided. Self finishes (ii) comprise the use of materials which are both structural and decorative. Examples of both these are:

Applied finishes:	Plaster.
	Dry lining in plasterboard.
	Wood panelling, etc.
Self finished:	Fair face brickwork using facing or other decorative or dimensionally accurate bricks.
	Fair face blockwork.
	Fair face concrete.

11.01
Finishes are either 'wet' or 'dry', depending on their composition.

'Wet' finishes include plaster, external rendering, external rough cast and pebbledash renderings, tyrolean finish, etc.

'Dry' finishes include timber boarding, tile and slate hanging, sheet plastic or grp, asbestos tiles and sheets, etc.

11.02
Internally, it is usual to provide finishes which either

(i) express the structure and provide a strong decorative texture such as fair face brickwork, or

(ii) provide a smooth hygienic surface suitable to take paint or similar applied finishes, e.g. gypsum plaster.

This module deals with finishes which fall generally into section (ii) and covers the application of plaster, dry lining techniques and decorative finishes to walls and ceilings and screeds and decorative finishes to floors. In addition, glass, its selection and glazing are included to complete the module.

12.00 Wall plastering, dry lining and external rendering
Plaster as a wet finish is applied to wall surfaces for a number of reasons:

(i) To level up uneven structural or partition walls to an acceptably flat finish.

(ii) To provide a surface suitable for the application of paints and decorative wall coverings such as wallpaper.

(iii) To provide a surface level enough to form a ground for thin ceramic wall tiles fixed by means of an adhesive.

(iv) To provide a reasonably close textured surface capable of receiving treatment such as gloss paint to render the surface easy to clean and consequently hygienic.

12.01
The surface or backgrounds for plastering are very varied and all present their own particular problems of adhesion, suction and movement. A wide range of plastering techniques and materials is available to meet these specific contingencies, especially as some of these are created by the ever increasing efforts to speed up the rate of building.

The three main groups of materials available and used for plastering are lime, portland cement and gypsum plasters and the general principles and techniques of their use are set out in CP 211:1966 *Internal Plastering*. All these materials are easily damaged by moisture and care must be exercised in their storage and handling. Plasters delivered to site must be stored in clean dry covered stores. Sand must be kept clean and dry and free from contaminating mud. Water must be clean main water.

12.02
Plastering is normally carried out immediately after first fixings and services carcassing have been com-

pleted. All door linings, frames and plastering grounds for fixing architraves and skirtings must be fixed plumb and level to act as datum for the final plaster surface finish by projecting the full thickness of the plaster.

All solid backgrounds such as brick or block walling must be free from dust, powdery efflorescence or loose and projecting mortar blobs. All electrical cables must be set in protective conduit, chased flush into grooves cut into the walls for that purpose.

12.03

Plaster is a relatively rigid material when set. Any movement in the background caused by shrinkage or thermal movement is transmitted to the plaster through its adhesion, causing cracking. This is very apparent when structural materials such as brickwork, blockwork or concrete adjoin.

The differential movement between the materials must be offset by preventing the adhesion of the overlying plaster to these materials where they come together. Expanded metal lath to BS 1369: 1947 should be fixed over thick building paper to the areas of likely movement before any plastering is carried out to prevent adhesion to the differing materials and at the same time provide sufficient key to carry the weight of the plaster.

12.04

External corners of walls (arrises) are very prone to mechanical damage, not only during the progress of the works but also in subsequent use. A method of preventing this is to fix an expanded metal bead to the corner, the sharp angle of which will provide extra resistance to damage and protect the surrounding plaster.

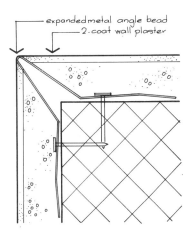

12.05

Accurate proportioning of materials and absolute cleanliness are essential for good plastering. Measuring boxes should be employed, used on a clean boarded platform (banker), the materials being mixed first dry and again after water is added. Boxes and board must be well washed between each batch.

Old mixes must not be used as a basis for new and once plaster has commenced to set, extra water must not be added. Mixers are often used when continuous production is required and must be washed out after each batch to ensure that no old material is incorporated in the new.

12.06

There are two basic types of plaster used to finish solid wall surfaces:

(i) Lightweight premixed retarded hemi-hydrate plaster to BS 1191 Pt. 2: 1973, produced from gypsum incorporating aggregates such as vermiculite or perlite. The advantage of premixed plasters is that they require only the addition of clean water and can be mixed indoors in cold weather to prevent freezing. The plaster is applied in two coats, the undercoat 11 mm thick being finished with 2 mm of neat finishing coat. Undercoats are produced to suit different surface backgrounds, the finish suits all grades. The final finish to walls plastered in this material is smooth with a matt, slightly mealy appearance.

12.07

(ii) Two-coat plaster comprises a rendering coat of either a proprietory slow setting browning undercoat or 1:1:6 or 1:2:9 cement/lime/sand rendering coats 11 mm thick finished with 3 mm neat anhydrous gypsum plaster (Sirapite) conforming to BS 1191 Pt. 1: 1973 Class C. The backing coat must be allowed to dry thoroughly before the setting coat is applied otherwise surface cracks will develop due to continuing shrinkage. The surface of this plaster is smooth.

An alternative to Sirapite finish is the use of a retarded hemi-hydrate gypsum plaster, usually known as Thistle plaster, but only when browning undercoat is used. Thistle plaster must conform to BS 1191 Pt. 1: 1973 Class B.

12.08

Where partitions are constructed of timber studding, these will usually be finished with plasterboard lath and skim coat plaster in the same way as ceilings are plastered (14.04 *et seq*). However, in some circumstances expanded metal lathing is used and this is strained as rigidly as possible over the timbers and well secured with galvanised staples.

Plastering is carried out in two-coat work, time being allowed to dry and shrink before the next is applied. Special semi-hydrate plasters containing hair and light weight metal lathing plasters are used.

12.09

Some backgrounds have insufficient suction to allow the plaster to adhere to their surface. To improve this the surface may be either

(i) hacked with a mechanical hammer to roughen the surface and provide a key for the plaster, or

(ii) covered with a material known as a bonding coat which provides additional suction.

12.10

Good plastering, when finished, should feel smooth under the hand, free from rough patches, dents, blisters and other defects. The quality of the finished work depends on the care and skill with which the work has been carried out. Backing coats should be applied thinly with a firm pressure built up to the thickness, brought to the required level with a straight edge and well scratched to provide a good key for the finish.

Finishing coats are applied in one coat and should not be over polished or they dust up. Arrises should be slightly rounded and the whole of the plastering allowed to dry out thoroughly before decorating.

12.11

To avoid certain cracking between plastered walls and ceilings it is often an advantage to fit a gypsum plaster cored cove into the angle. The cove is self finished and, after cutting to length and mitring the angles, the cove is nailed to joists and screwed to wall plugs and the joints filled with cove adhesive.

12.12

Walls are often finished with a dry lining technique incorporating 9.5 mm aerated gypsum plasterboard manufactured in accordance with BS 1230: 1970. The plasterboard requires a level and true surface for fixing and this is provided by fixing small impregnated fibreboard pads to the walls with Carlite bonding at specific centres, care being taken to bridge the joints

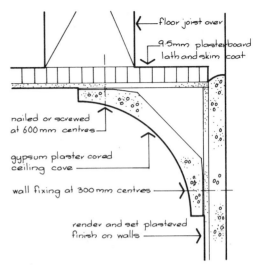

Detail of 'Gyproc' plaster cored ceiling cove

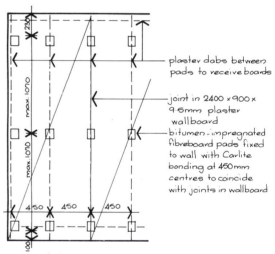

Fixing layout for drylining masonry wall by 'Thistlebond' method

between adjacent boards. Dabs of plaster applied to the pads enables the pre-cut plasterboard panels to be secured to the wall by suction as the plaster sets.

12.13

Three profiles of plasterboard are used for this finishing technique:

(i) Taper edge from which a flush seamless finish can be obtained by tapering and filling the shallow well formed by the edges.

(ii) Square edge which may be close butted and finished with a cover strip.

(iii) Bevelled edge where a vee joint becomes a feature of the wall.

All three board types can be obtained with an aluminium foil backing which, when the joints are taped with a self-adhesive foil strip, will act as a vapour barrier (see *Building Technology 1*, 16.09 and 22.15) (14.03 Grades and thicknesses).

12.14

External renderings are applied to the walls of buildings for a variety of reasons:

(i) Aesthetic, to provide a particular effect to the external appearance.

(ii) Protective, to protect the materials of which the wall is constructed from rain and frost and improve its weathering characteristics.

(iii) Economic, to provide a quality finish over second quality materials used in the structural wall.

External renderings must be porous, and provided with an open textured finish if they are to successfully withstand the chemical and weathering problems to which they will be subjected. Weak mortar mixes based on lime with a wood floated or steel scraper finish, while tending to absorb more moisture in wet weather, assist in expediting drying out, eliminating the problem of entrapped water behind the rendering. In addition, the tendency of a stronger mix to crack is entirely absent in weak lime mixes.

Adhesion is a problem. Hard dense backgrounds such as concrete need either a splatter dash coat of cement and sand (1:2) cast onto the surface or the application of a proprietory bonding coat to improve adhesion.

12.15*

The joints of brickwork should be well raked out to a depth of 19 mm for rendering to improve the key.

Materials should conform to relevant BS proportioned by volume as for plastering and either mixed by hand or mechanically. A mix of 1:1:6 (cement/lime/sand) is a good reliable guide, an undercoating first being applied 13 mm thick with a trowel and well scratched to provide a key for the finishing coat. Free bottom edges of the rendering should be bellied out 25 mm for a height of 100 mm and undercut to throw water clear of the wall below.

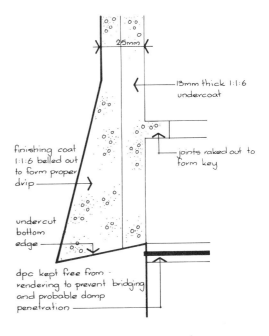

Care must be taken not to bridge the d.p.c. by rendering.

The second coat should be about 10 mm thick, a smooth finish being given by the use of a wood float, felt faced floats tending to provide a rougher texture by picking up the surface. Using a steel scraper on a wood float finish produces a natural open texture finish, and this can be accentuated with an expanded metal faced float.

12.16*

A satisfactory rendering technique for rural building is known as Tyrolean. Unfortunately, this coarse finish is not satisfactory in urban areas as it attracts and holds dirt and grime. The preparation of the background is similar to the first rendering described, following a two coat principle.

The application of the Tyrolean rendering coat is by means of a hand-powered machine which throws the second coat material onto the backing, building up the thickness to a textured coat of granular consistency. Windows, doors and painted surfaces require masking while this work is being carried out.

13.00 Ceramic wall tiling

Ceramic wall tiles to BS 1281: 1966 are used in large quantities to provide a wall surface that is both hygienic and durable. The tiles are manufactured by applying a liquid glazing material onto a clay based 'biscuit' and firing this at high temperature in a kiln. A wide variety of colours and applied patterns are available but are generally available only in the following sizes and thicknesses:

152 x 152 x 5 mm
152 x 152 x 6 mm
108 x 108 x 4 mm

The edges of the tiles are rounded over (cushioned) to improve jointing and provided with spacer lugs to ensure even and regular joint widths of 3 mm. In addition, the spacers act as 'crushing points' which will fail without breaking the surface of the tile if the backing wall shrinks after the tiles have been fixed.

13.01

The larger size of tile can be provided with a great variety of fittings to carry out complicated tiling work, angle tiles, coves and cappings being the most common in use. Glazed earthenware fittings for soap dishes and toilet roll holders can be obtained in matching sizes and colours.

13.02

Ceramic tiles can be fixed to a wide range of backgrounds. Where tiling is to be fixed to brick or block walls, these must be perfectly dry to ensure that all shrinkage which might affect the tiled surface, has taken place. The background for the tiles may be either:

(i) a 1:4 cement and sand rendering coat 13 mm thick floated level to receive the tiles, or

(ii) a plastered background to which the tiles are fixed.

13.03

The ceramic tiles are fixed by one of three methods depending on the use or abuse the tiled surface will receive:

(i) The 'thin bed' method for normal dry interior installations onto either a rendered or plastered background. In this a thin floated coat of suitable adhesive 3 mm thick is spread over an area of about $1m^2$ and carefully ribbed with a notched trowel. The dry tiles are then pressed into position, taking care with square edge tiles to leave a joint of at least 1.5 mm.

(ii) The 'solid bed' method used for areas where water may be a problem, e.g. showers. This method requires a rendering coat backing and the adhesive used is left solid to receive the tiles.

(iii) The 'thick bed' method is used when ceramic tiles are to be fixed to an uneven wall surface, e.g. fair face brickwork and where a cement and sand rendering is either not acceptable or would be too unstable to ensure proper adhesion. A setting bed of a proprietory cement or rubber based adhesive 13 mm thick is applied to the wall surface, and the tiles fixed as described for the 'solid bed' method.

13.04

The open joints between the tiles are filled by grouting. To allow the adhesive to set properly, grouting should not be carried out less than 24 hours after fixing. As the grout should be resilient, a proprietory expanding grout should be used to provide a suitable joint. The grout is pressed firmly into the joints and around the edges of the tiles, and the excess wiped off before the tiling is polished up with a clean cloth.

The surface of the tiled area should be straight and level without irregularities or distortion. The fixing and workmanship employed should comply with CP 212 Pt. 1: 1963 *Wall tiling*.

14.00 Ceiling plastering and finishes

The material we know today as plaster has been in use in the UK for the finishing and fireproofing of ceilings for much longer than the general application to walls. Originally used to provide a degree of fire resistance to the underside of thatched roofs, the craft of plastering developed great artistic and technical merit during the 17th and 18th centuries.

The skill became commercialised in the 19th century and lost much of its beauty and freshness. Today ceilings are merely flat undecorated areas of plaster provided a background for flat paint or decorative paper.

14.01

The base for plaster is a wall board made from gypsum encased in a paper covering. This board can be obtained in two sizes, 2440 x 1220 mm and 1220 x 406 mm. The smaller size is known as gypsum lath and is generally used for ceilings as it is easier to fix, being smaller and consequently lighter. In addition ceilings using this material do not suffer so much from shrinkage cracking.

14.02

Both materials are obtained in two grades, standard and insulating grades, the latter having an aluminium foil bonded to one face. The aluminium foil acts both as a vapour barrier (when the joints are sealed with a self adhesive aluminium strip (12.13)) and as a reflective insulant when fixed with the foil facing outwards.

14.03

Lath and wallboard are obtained in two thicknesses, the selection depending on the centres of support, 9.5 mm grade being satisfactory up to 400 mm centres, over this dimension and up to a maximum of 600 mm. 12.7 thickness material should be used.

14.04

Lath is easy to cut with a fine toothed saw or sharp knife and should be fixed across the supporting joints, laid breaking joint and fixed with sheradised plasterboard nails at 150 mm centres. The cut edges are butted together, a gap of 3 mm left between the rounded longitudinal edges.

14.05

Finishing the lath can be carried out in either of two ways depending on the quality of the work:

(i) By applying a setting coat of neat retarded hemihydrate board finish gypsym plaster.

(ii) By applying a bonding grade undercoat finished with a lightweight plaster to BS 1191 Pt. 2: 1973 *Premixed lightweight plasters*. This is a much superior finish as the material is slightly resilient which reduces cracking caused by structural movement.

With both methods the angles between the walls and ceilings should be scrimmed with 87 mm wide jute scrim cloth set in neat plaster to reduce the risk of perimeter cracking. The finishing plaster is mixed in a plastic bucket with clean water and trowelled smooth to a finished thickness of 3 mm.

14.06

Another ceiling finish generally used for domestic work is a patent material which is worked to a texture on application. A gypsum plaster lath background is provided but omitting the perimeter scrim. The finish is mixed on site from self-coloured materials and applied direct onto the surface of the lath, a strong texture being applied by using various float profiles.

15.00 Floor screeding and finishes

It is rarely possible to lay a concrete bed sufficiently level and smooth to enable thin floor finishes to be applied direct. Consequently oversite concrete forming solid floor construction is generally set down below the finished floor level to allow an intermediate layer of material to be applied both to level up the construction accurately and to provide a smooth bed to receive the floor finishes.

This intermediate layer is formed from a mixture of cement and sand usually in the proportions of 1:3 and is termed a screed.

15.01

Most screeds are laid on the 'separate construction' basis, i.e. after the oversite concrete has been allowed to harden. The surface of the concrete is usually roughened during construction by the consolidation process known as 'tamping' and to obtain a key for the screed a layer of neat cement grout (cement mixed with water) applied to the surface will usually provide adequate adhesion. When the subsoil is excessively wet or where the floor finish specifically requires a continuous horizontal d.p.c. to protect the floor from rising damp, two or three coats of a cold bitumen emulsion or a similar preparation may be used, the last coat being well blinded by casting onto the tacky surface dry coarse sand to provide a key for the following screed. Care must be taken to joint the d.p.c. in the floor to that provided in the walls to avoid rising damp bypassing both around the permeter of the building.

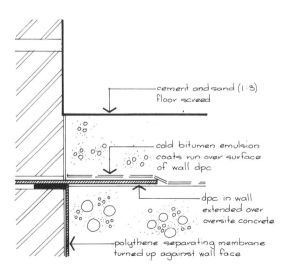

If the oversite concrete is too smooth to provide a proper key for the screed its surface must either be hacked with a mechanical hammer or treated with a proprietory bonding coat.

15.02

Dense cement and sand screeds should be laid in bays, the size of bay being directly related to the thickness of the screed. With screed thickness of 40 mm (the minimum thickness in practical use) the bay size should not exceed 15 m^2 and the ratio of sides near to 1:1½. Where expansion joints occur in structural slabs these should be continued through the screed.

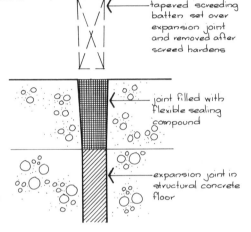

15.03

Screeds are mixed by hand, or in a mixer, and laid as dry as possible. The thickness is maintained by timber battens called 'rules' and the surface compacted and finished with a screeding board and finally with a steel trowel or float.

15.04

The screed must be cured and to effect this, drying must be controlled to make sure that the material gains initial strength before drying shrinkage which induces the risk of cracking and curling. Consequently the screed should be covered with polythene for at least seven days after laying. Thereafter the screed will continue to dry naturally at the rate of approximately 25 mm in thickness every four weeks. Accelerated drying by heaters should be avoided wherever possible.

15.05

Accelerated drying or lack of adhesion to the subfloor can cause hollow areas, detected by tapping the surface of the screed. These areas must be hacked up and relaid. Minor irregularities can be repaired by the application of a latex screed to level up the surface.

15.06

Floor finishes are provided for a number of reasons. These may be summarised as follows:

(i) Hygienic. To provide a surface which is impervious to dirt and liquids and which by simple washing and cleaning maintains a reasonably aseptic covering. Examples of flooring meeting this requirement are quarry tiles, vinyl tiles and sheet flooring.

(ii) Resistance to impact. To provide a surface which will withstand heavy loading from machines or rough wear from usage. Examples of flooring meeting this requirement are quarry tiles, granolithic pavings and some species of timber used for the manufacture of wood blocks such as Iroko and Gurjun.

(iii) Appearance. To provide a floor which has a pleasant aesthetic effect and which with regular attention will have a long and satisfactory life. Examples of flooring to meet this requirement are hardwood strip and mosaic, hardwood block flooring and cork tiling.

There are other types of flooring manufactured but these generally fall into the categories enumerated above.

15.07

Another distinguishing feature of various floor finishes is the manner in which they are laid. Each type of floor has its own peculiar problems and recommendations but these may be summarised into groups as follows:

(i) Thin inorganic materials (vinyl) and organic blocks (wood blocks and mosaic and cork tiles) which can be secured by setting in a specialised adhesive direct onto the screeded subfloor. No jointing is required as the separate units are set close to one another without gaps.

(ii) Thick organic strip material such as hardwood strip and softwood boards which need continuous control to counteract their natural tendancy to twist and warp. This is effected by nailing these floors down to dovetailed hardwood strips set into the concrete floor, the spaces blinded by a coat of bitumen before screeding. The boards are well

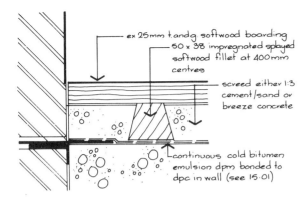

cramped up together and provided with tongued and grooved joints in a similar manner to softwood boarded floors. (see *Building Technology 1,* 21.10 (ii)).

(iii) Thick blocks or tiles of fired clay or ceramic material (quarry tiles) which have great resistance to shrinkage and must be bedded in a 10 mm thick 1:4 cement and sand bedding, separated from the surface concrete by a layer of building paper or polythene. The tiles are provided with 6 mm joints which are both decorative and take up the irregularities in the tile itself.

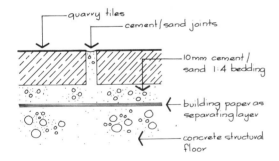

(iv) Jointless flooring either prepared cold from a mixture of cement and organic aggregate such as granite chippings (granolithic paving) or by melting and applying organic material hot direct to the subfloor (asphalt flooring). Both types of flooring are applied direct to the oversite concrete and both require good adhesion to be successful.

15.08
The physical dimensions of the different floor finishes are important as these will determine the level of the screed or concrete subfloor.

Vinyl tiles   305 x 305 x 1.6 mm thick (min)
Vinyl sheet   2.4 m wide x 1.6 mm thick (min)
(BRS Digest 33 (2nd Series). BS 3261: 1960)

Hardwood blocks   228 x 89 mm x 25 mm (nominal)
                   305 x 89 mm x 25 mm (nominal)
(CP 201: Pt. 2: 1972. BS 1189: 1959)

Hardwood mosaic   460 x 460 x 13 mm thick (nominal)
(CP 201: Pt. 2: 1972. BS 1187: 1959)

Cork tiles   305 x 305 x 4.8 mm thick (min)
(CP 203 Pt. 2: 1972)

Hardwood strip   16, 19, 21 and 28 mm in thickness, widths and lengths as obtainable. (Batten widths under 100 mm wide are preferable).
(CP 201 Pt. 2: 1972)

Softwood boarding as above for hardwood (BS 1297: 1970)

Clay quarry tiles   Type A tiles have greater dimensional unevenness but are more attractive when laid. Type B tiles have a fine smooth even face and texture and are manufactured to closer tolerances. Type A, various sizes up to 225 x 225 x 32 mm. Type B, up to 150 x 150 x 13 mm (max).
(CP 202: 1972. BS 1286: 1945. BRS digest 79 2nd series)

Granolithic paving   Bays not exceeding 15 m² by 40 mm thick minimum (bay separation by means of 6 mm thick ebonite strips).
(CP 204 Pt. 2: 1970)

Asphalt flooring   15 mm thick in one layer for domestic and other light wear installations.
(CP 204 Pt. 2: 1970. BS 1162, 1418, 1410: 1966)

15.09
Expansion joints to allow movement in the concrete subfloor are necessary when floor finishes of rigid materials such as wood blocks and quarry tiles are used. These expansion joints formed from a water resistance compressible material are either

(i) provided around the perimeter of the room between the floor finish and the coved skirting (quarry tiles), or

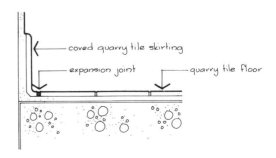

(ii) set under the skirting where it can be concealed (quarry tiles), or

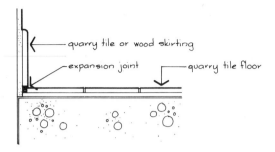

(iii) between the straight joint margin usually provided to wood block flooring and the herringbone field.

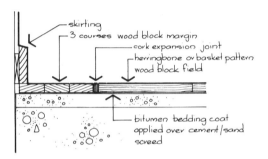

In addition the expansion joint should be provided across door openings. It is usual to provide this centrally under the door where two differing floor finishes meet.

Hardwood block and strip floors and cork tiles are cleaned off by a mechanical sanding machine after laying to provide a smooth level surface. The sanded surface is then filled where indents make it necessary, sealed with a proprietory seal and either wax polished or a further coat of seal applied to provide a suitable wearing surface to the material.

15.10

Care must be taken with quarry tile and granolithic pavings to cure and protect these floors in the same way as floor screed are protected (15.04).

Wood and thin material floors also need protection from damage and abrasion from finishing trades after laying by the application of building paper or a layer of sawdust.

16.00 Applied decoration and painting

The painting of building structures is carried out for three reasons:

(i) To protect the structure or substrata.

(ii) To enhance its appearance.

(iii) To seal and provide a hygienic and easily cleaned surface.

16.01

Various features reduce the effectiveness of paints depending on the physical formation, the degree of flexibility of the applied paint film and the chemical composition of both substrata and paint. These can be summed up as follows:

(i) Moisture content and movement. Timber must have a moisture content less than 25% and for building purposes this should be between 15 and 18% for exterior and 10–12% for interior use. Otherwise the excess moisture content will cause blistering of the paint film and opening up of mechanical joints forming points for water penetration.

(ii) Chemical reaction of the substrata. Examples of this are rusting of ferrous metals, efflorescence caused by leaking of salts in brickwork and cementitious backings and sulphate attack.

(iii) Exudation of resinous matter. Most timbers, hardwoods and softwoods exude resinous material and oils which can interfere with the drying, hardening and adhesion of paint. While kiln drying has the effect of lowering the resin content of some timbers, in pines, e.g. Columbian and Oregon, the exudation occurs at knots, pitch streaks and pockets. Some hardwoods such as afromosia, gurjun, agba and keruing (all used for sills) suffer in a similar way.

(iv) External influences. Interstitial condensation can cause paint failure, natural rainfall can gain entry at the bottom joints of window frames and doors, the end grain of door posts and exterior boarding and on contact with masonry. Strong sunlight can cause a degree of colour fugitiveness in certain pigments such as blue and dark grey.

16.02

To ensure satisfactory and durable finish the paint must be prepared and applied in accordance with certain standards and also in accordance with the particular recommendations of the paint manufacturer.

Mixing is important, thorough integration of all ingredients prior to use is essential. Paint in store must be protected from excess heat and cold to prevent deterioration in the container. Any thinning required must be done with the recommended thinners in the proportions described by the manufacturer.

Application of primer should always be by brush, subsequent coats being by brush, spray or roller as appropriate to the surface to be covered. Each coat of paint must be hard dry before the next is applied and all surfaces must be thoroughly dry before any paint is applied. Each coat except the last should be lightly rubbed down with fine glass paper before the next is applied. Painting should not be carried out in driving rain, fog or frosty weather or in direct sunlight.

16.03

The problem of resinous exudations from knots and pitch pockets common in some timbers is contained by treating the affected areas by two coats of a preparation known as 'knotting'. This is a proprietory material made by dissolving shellac in methylated spirits, each coat being allowed to dry before the next is applied (1.09).

Where the whole surface of the timber is likely to exude resinous material (e.g. oak) two coats of an aluminium leaf primer sealer should be applied to the whole surface of the timber member (1.09).

16.04

A good sound primer is an essential base for a satisfactory and lasting paint system. So far as timber and woodwork is concerned, this should, in conjunction with the application of knotting, be carried out in the joiner's shop. The primer specified should match in manufacture the remainder of the paint system used. The following primers are satisfactory for the substrata indicated:

(i) Softwood. Pink primer containing a high proportion of white lead plus some red lead formulated to BS 2521 has good filling properties, reasonable durability to withstand exposure and excellent for external use. Primers with an oleo-resinous or alkyl base are satisfactory for internal use. Emulsion primers may be used for internal work being quick drying and non-toxic.

(ii) Hardwood. Aluminium based primers are most satisfactory being good fillers of open grain and having a high resistance to water penetration they are best used on sills and the backs of frames in contact with brickwork or other masonry surfaces (1.09).

(iii) Ferrous metals. After removal of rust, dirt, oil and moisture a primer having filling and rust inhibiting qualities should be applied. The best is probably red lead primer to BS 2523 although zinc chromate primer and other zinc rich primers are satisfactory where the surface is in good condition and the exposure not severe (see *Building Technology 1,* 23.07).

(iv) Galvanised surfaces. For galvanised metal windows and similar situations a calcium plumbate primer should be applied to all surfaces before they are glazed. Beads, if provided, should be removed and painted separately (1.09).

(v) Alkali prone surfaces. Certain materials such as cement and sand renderings, insulating boards and asbestos sheets are alkaline and must be sealed with an alkali-resisting primer if they are to be finished in a gloss paint. With an emulsion finish this precaution is not necessary.

16.05

Defects and nail holes need to be filled before the application of further paint coats. Linseed oil putty shrinks and needs 48 hours to harden sufficiently before being able to take paint and should be avoided. Proprietory fillers for external use can be obtained.

16.06

Primers become powdery if left too long exposed to the elements before the application of protective paint coats. If this has occurred the primer must be scraped down and a further coat applied. The success of a paint system depends largely on the quality of the primer.

Undercoats must be of the same manufacture as the finishing coat. Generally one undercoat plus a coat of gloss finish is the minimum specification but for better class work and a superior finish two undercoats are applied. Where the paint is formulated and sold as suitable for 'gloss on gloss' application a second coat of gloss may be applied. This will greatly prolong the life of the paintwork, the undercoat providing the pigment body and the gloss the protection.

Gloss finish provides the weatherproof seal to the underlying pigmented undercoat. While one coat is sufficient for internal work a second 'gloss on gloss' application will ensure total coverage and a greatly extended life to the paint system.

16.07

Water bound paints or emulsions are used in large quantity for the decoration of internal walls and ceilings. As these paints are porous and have little 'body' careful preparation of the plaster is required. This should be lightly rubbed down with fine sandpaper, cracks and holes filled with an expanding filler. If the surface is very dry or has high absorbancy a sealing coat of thinned emulsion paint should be first applied, followed by two further full coats straight from the container.

17.00 Glass and glazing

The glazing of domestic structures is normally carried out when the roof is weathered and as the scaffolding is being struck. This enables the building to be made weather tight.

Glass for glazing is manufactured to BS 952: 1964, *Classification of glass for glazing*, and four types are in general use for building purposes:

17.01

(i) *Sheet glass* is the cheapest and is obtained in a number of thicknesses and sizes:

Nom. thickness (mm)	Nom. max. size (mm)
3.0	2030 x 1220
4.0	2030 x 1220
5.0	4.65m² (max dim 2640)
5.5	9.30m² (max dim 2640)
6.0	9.30m² (max dim 2640)

This glass is never perfectly flat but is quite satisfactory for housing.
Three qualities are available:

OQ – Ordinary glazing quality. For general building work.
SQ – Selected glazing quality. For good class building work.
SSQ – Special selected glazing quality. For cabinet work.

17.02

(ii) *Translucent glasses* ('obscured' or 'patterned') are rolled glasses usually with one flat and one textured surface. The degree of obscuration varies from low to high (a–e). These glasses are obtained in sheets varying in size from 1270 x 1520 mm to 1320 x 2130 mm and in two thicknesses, 3 and 3.5 mm, varying with the pattern of texture.

Commonly used translucent glasses are as follows:

Pattern	Decree of Obscuration	Thickness (mm)
GROUP 1		
Arctic large	b	3
Flemish small	a	3
Pacific	c	3.5
Stippolite	c	3.5
GROUP 2		
Broad reeded	a	3.5
Narrow reeded	a	3.5
Festival	c	3
Spotlyte	c	3.5
GROUP 3		
Plain Cathedral	a	3

17.03

(iii) *Float glass* which has largely replaced polished plate glass, due to its lower cost and which can be obtained in a variety of thicknesses and sizes as the following examples:

Nom. thickness (mm)	Nom. max. size (mm)
3	1270 x 1270
5	2540 x 2280
6	4650 x 3170
10	7110 x 3300

Float glass is flat, parallel and fine polished giving clear undisturbed vision. Two qualities are available for building purposes:

GG – Glazing glass. For general purposes
SG – Selected glazing quality. For better class work such as shop windows.

17.04

(iv) *Wired glass* used in areas of special glazing to protect the occupants of the building from the perils of shattered glass, e.g. glazed lights in doors and roofs and fire resisting construction. Two types of glass are available:

(a) Georgian polished wired glass which is a polished plate glass reinforced with a 13 mm square wire mesh. This glass is 6 mm thick and obtained in sheets 3300 x 1830 mm maximum.

(b) Georgian wired cast glass, a translucent rough cast glass with similar wire mesh reinforcement, 6 mm thick and obtained in sheets 3710 x 1830 mm maximum.

17.05

In addition to ordinary glass, the glazing of windows is often carried out by using two sheets of glass, usually 3 mm thick float glass separated and with the intermediate air exhausted by vacuum. Separation is achieved either

(i) by metallic spacers which are cemented to the glass, or

(ii) by sealing the edges of the glass together during manufacture to produce a hollow wholly glass construction.

The use of double-glazed units in windows and doors greatly increases the thermal resistance of the glazing and must be considered obligatory if the windows are of any size to enable the external fabric of the building to conform to the requirements of the Building Regulations in respect of thermal resistance (F4).

17.06

Glass must be secured into the window or door to prevent weather penetration around its perimeter or dislodgment by pressure or vibration. This is effected in a number of ways, as follows:

(i) Timber joinery

(a) Windows. By the use of linseed oil putty together with a small fixing nail or sprig to provide mechanical fixing.

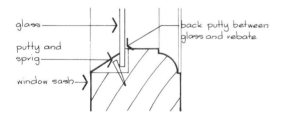

(b) Doors. By bedding the glass in mastic or washleather and securing the glass with a hardwood bead fixed with brass screws and sockets at about 225 mm centres.

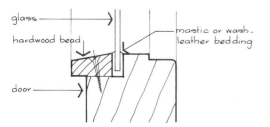

(ii) Metal windows and doors

(a) External glazing. By securing the glass to the frame with a wire clip and the use of a metallic glazing compound (similar to linseed oil putty but specially formulated for glazing to metal)

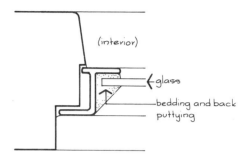

Detail of external glazing using metal glazing compound

(b) Internal glazing. By bedding the glass in metallic glazing compound and securing with galvanised pressed steel beads secured to the metal section by means of stainless steel self tapping screws or patent clips.

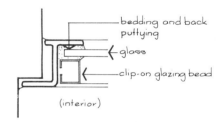

Detail of external glazing using spring clip glazing beads internally for fixing

Special glazing techniques are used for double glazing depending on the type of unit used by the particular manufacturer.

17.07

Care must be taken to ensure that the joint between the glass and the back of the glazing rebate is solidly back-puttied to avoid the glass rattling under wind pressure, suction extracting the glass from the light or water penetration in bad weather. All large and heavy sheets of glass should be bedded onto nylon spacers to support the weight.

Module D Second Fixing Joinery and Doors

18.00 Second fixing joinery — fixing and protection

We have seen that first fixing joinery is installed in the building structure for a number of reasons not repeated here (1.00).

When the services and wet finishing trades have been completed second fixing joinery can then be installed to provide:

(i) Complete enclosure, security and protection by hanging the external doors to the frames fixed earlier.

(ii) The fixing of trim, known as architraves, to the perimeter of internal door frames and linings to cover (master) the exposed joint between the woodwork and the plaster.

(iii) The fixing of trim, known as skirtings, to the perimeter of rooms at floor level to protect the plaster at this point from damage during floor cleaning and then master the joint between plaster and floor finish.

(iv) The complete enclosure of internal spaces by hanging the internal doors to the frames and linings fixed earlier.

Completion of this work will provide a clear field for decorators and floor layers to move into the building and effect its completion.

18.01

Immediately the walls and ceilings have been plastered it is possible for the electrical contractor to return to site and complete the installation. This involves providing and connecting switch plates, socket-outlets, ceiling roses with pendant drops and bayonet sockets and similar items to the cables protruding from the installed boxes in walls and ceilings.

These items are loose fixed to enable them to be temporarily removed by the decorators for paint and wallpaper to be applied behind the cover plates. In addition the plumber and heating engineer can commence carcassing work at this stage.

18.02

The installation of second fixing joinery follows accepted trade practices utilising standard items of fixings and ironmongery developed for the purpose.

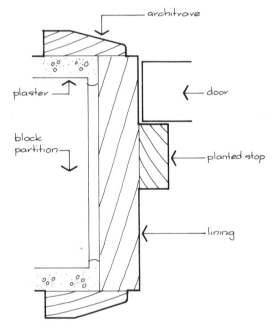

Detail showing architraves fixed in position to internal door lining - used in conjunction with painted softwood linings and trim

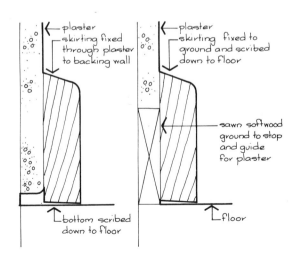

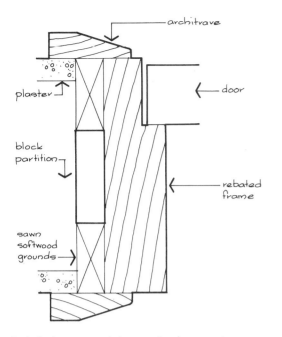

Detail showing architraves fixed in position to sawn softwood grounds - used in conjunction with clear-finished hardwood frame and trim

nails to either the frame or lining or to prepared timber grounds. Round lost head wire nails are usually employed for the purpose, well punched home and the holes filled in before the work is decorated. In cheap work where grounds are not provided, the skirtings are usually nailed through the plaster into the lightweight block walling with 65 mm cut steel nails, of a pattern generally used for securing softwood flooring.

18.03

The fixing (hanging) of doors depends on a number of factors as to the method selected and the material to be employed. These may be summarised as follows:

(i) The overall width and height of the door(s).

(ii) The thickness of the door.

(iii) The weight of the door.

(iv) The constructional framing of the door.

(v) The construction and strength of the frame or lining (if provided).

(vi) The designed opening of the door in relationship to the frame.

All these must be taken into consideration when selecting the hinges on which most doors are hung. Exceptions will be dealt with separately.

18.04

There are two principal forms of hinge used for hanging doors for domestic purposes:

(i) The *butt* hinge which comprises two separate flaps, each provided with a knuckle connected by a pin. These hinges are made in a variety of materials as follows:

(a) Pressed steel for light pattern doors.
(b) Steel (single and double) for heavier pattern doors.
(c) Cast and welded iron for heavy doors, especially external doors.
(d) Brass, either pressed for light internal doors *or* solid drawn for heavier patterns.
(e) Nylon for light pattern internal doors.

Three patterns of butt hinge are in general use:

Standard pattern where the pin is set on the centre line of the flaps.

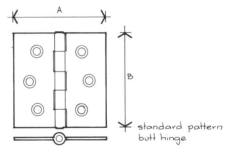

Cranked pattern where the pin is offset to one face of the flaps.

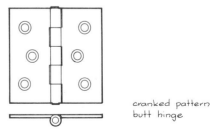

Extended or parliament pattern where the flaps are extended and lengthened to enable the door to be set to pass an obstruction when open, such as an externally opening door required to open through 180°.

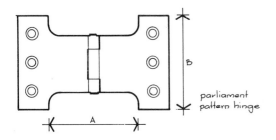

parliament pattern hinge

(ii) The *strap or tee* hinge usually comprising a narrow flap for screwing to the frame secured by a knuckle or pin to a long tail provided to enable the maximum width of door to be directly supported on the hinge.

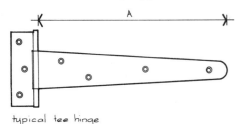

typical tee hinge

The hinges are made in the following principal materials:

Pressed steel for doors of light weight.
Malleable iron for doors of medium weight up to about 1500 mm wide.
Cast iron for very heavy doors up to 1750/1800 mm wide.

Two varieties of the tee hinge are in common use:

(a) Where the knuckle is replaced by a heavy hook and ride, designed to support the weight of a very heavy door. A good example is the Collinge hinge made in cast iron and obtainable up to a tail length of 1524 mm.

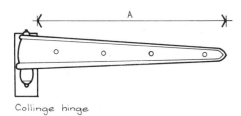

Collinge hinge

(b) Alternatively where no frame is provided, the flap may be replaced by a hook with a long tail for building directly into a brick or masonry reveal.

typical hook for carrying end of hinge where no frame is provided

18.05

There are a great many varieties of hinge produced from these two basic types, specially designed for specific purposes. Examples are as follows:

(i) Rising butt hinges which raise a door automatically above the floor level when the door is opened to clear carpets and other loose floor coverings. These hinges ensure that a door is self closing, although they are not permitted in this connection for fitting to fire doors by most fire prevention authorities.

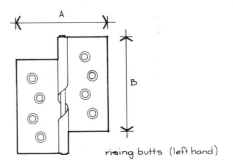

rising butts (left hand)

(ii) Continuous hinges known as piano hinges which are used for quality joinery where a thin door of some weight needs continuous fixing and support throughout its length.

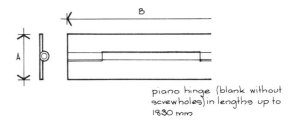

piano hinge (blank without screwholes) in lengths up to 1830 mm

(iii) Invisible hinges fixed completely within the frame rebate and used for doors concealed in continuous panelling and similar situations.

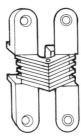

invisible hinge with solid brass joints

(iv) Cranked hinges for windows and door provided with stormproof face beads which would obstruct the operation of a normal hinge.

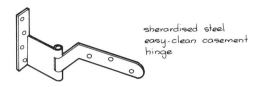

sherardised steel easy-clean casement hinge

In addition there are a great number of special door hanging devices applicable to highly specialised doors which will be included in the text in the appropriate sections.

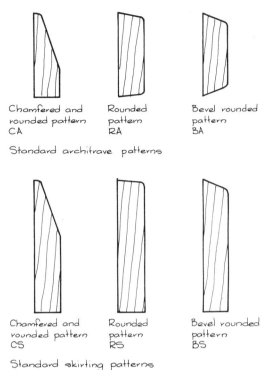

Chamfered and rounded pattern
CA

Rounded pattern
RA

Bevel rounded pattern
BA

Standard architrave patterns

Chamfered and rounded pattern
CS

Rounded pattern
RS

Bevel rounded pattern
BS

Standard skirting patterns

18.06
The selection, preparation and protection of all second fixing joinery should follow the recommendations described in Module A (1.00 et seq).

19.00 Architraves and skirtings*
These items fall under the name of 'trim' and are usually run out by machine in long random length from which those appropriate for fitting and cutting with the minimum waste are selected for particular situations. In general the wood selected follows that used for the remainder of the joinery although sections with long straight grain and an absence of knots and pockets are usually preferred.

19.01
In good class work the sizes and profiles of architraves and skirtings are run out to match the architect's details, but where the joinery is of standard stock quality (i.e. produced in bulk to standard sections and overall sizes), trim is used conforming to BS 584: 1967 Wood trim (softwood). This provides for a number of standard profiles and sizes from which the following are selected:

(i) *Architraves*
   (a)   Ref No. 14 CA 45   T X W (mm) 14 x 45
                20 CA 70   10 x 70
   (b)   14 RA 45   14 x 45
                20 RA 70   20 x 70
   (c)   14 BA 45   14 x 45
                20 BA 70   20 x 70

(ii) *Skirtings*
   (a)   14 CS 70   14 x 70, 95, 120
                20 CS 70   20 x 70, 95, 120
   (b)   14 RS 70   14 x 70, 95, 120
                20 RS 70   20 x 70, 95, 120
   (c)   14 BS 70   14 x 70, 95, 120
                20 BS 70   20 x 70, 95, 120

19.02
Standard stock joinery trim is usually jointed at the corners by mitres. This is usually satisfactory so long as the moisture content of the wood is correct at the time of installation and the humidity and temperature of the building is closely controlled. Otherwise shrinkage takes place and the joints open up.

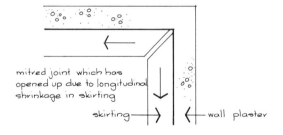

mitred joint which has opened up due to longitudinal shrinkage in skirting

skirting

wall plaster

19.03
In good quality work the joints are varied to suit the position occupied.

External angles of both skirtings and architraves are mitred as before but internal angles have one section scribed over the next. This ensures that if the longer length shrinks (as it may well do) the joint slides and does not appear as an open joint.

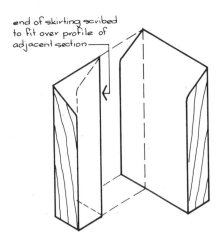

end of skirting scribed to fit over profile of adjacent section

In addition, the bottom edge of the skirting is usually scribed down right to the floor finish to effect a tight joint. This is rarely carried out in cheaper quality work (18.02).

19.04

All surfaces of softwood trim should be properly knotted and primed before being fixed. Where hardwood trim is provided or the timber is to be left natural and protected by a clear finish, the bedding face of trim should be protected by applying a coat of an aluminium primer sealer to avoid any leaking of salts from the wall surface into the wood.

20.00 Function and construction of basic doors

Doors are installed in buildings for a number of reasons:

(i) To protect the interior of a building from a hostile external environment, i.e. from rain, snow and variations in temperature.

(ii) To provide security to the interior of a building and its contents.

(iii) To protect the various compartments of a building and its occupants from the hazards of fire and noxious fumes.

(iv) To provide localised security to particular rooms or areas, protecting their contents from unauthorised use or damage.

(v) To provide privacy to the occupants of individual rooms or areas.

(vi) To provide a degree of sound reduction between separated areas of a building in conjunction with the frame and wall construction.

In addition one must include the obvious function of a door to provide access not only into the building but also into the various rooms and compartments into which it is divided.

20.01

Doors must be constructed to fulfil a number of conditions if they are to function effectively throughout the life of the building.

(i) They must be capable of being connected (hung) to the prepared opening in such a manner as to not only open and close easily and effectively but also to do so without continuous adjustment.

(ii) They must be so constructed as to fulfil their performance requirements in respect of weather exclusion, fire resistance, sound reduction or any other reasonable requirement.

(iii) They must be so constructed as to maintain their shape with the minimum of movement or distortion.

20.02

Timber for doors should be 'door stock quality' free from defects listed in BS 1186 (1.04 *et seq*), and selected from species suitable for the proposed situation. Moisture content should be carefully controlled and joints should permit movement caused by variations in temperature or humidity without exposing open joints.

20.03

Doors are flat areas, usually principally constructed of wood or wood based manufactured panels in which the length and width greatly exceeds the thickness.

Traditionally, doors were made up (constructed) from a number of wide or narrow boards, their long edges accurately finished and, perhaps, moulded to provide decoration and concealment to the joint which is made by means of a loose tongue inserted in grooves cut in the edge and well glued up.

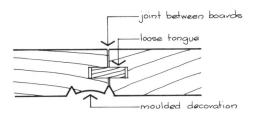

joint between boards
loose tongue
moulded decoration

A development of this jointing technique was the provision of the tongued and grooved joint (t. and g.) incorporating either a vee joint or bead moulding which was left unglued to allow the boards to shrink in use as the moisture content of the wood was reduced by the atmosphere to its optimum.

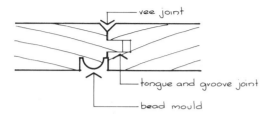

To reduce the shrinkage, and perhaps control it, and to give horizontal stability to the whole area, horizontal timber members called ledges were secured to the back of the door, originally with nails but now by stout gauge short countersunk screws, often set in slots to allow for movement in the timber ledge.

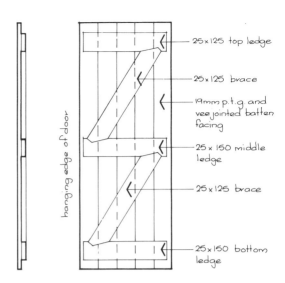

Internal elevation of a softwood ledged, braced and battened door

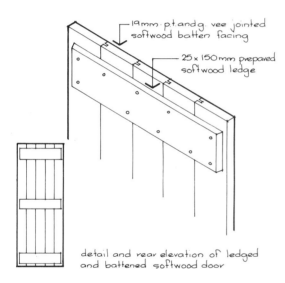

detail and rear elevation of ledged and battened softwood door

A door constructed in this way allows all members to take up movement properly without stress or undue distortion.

20.04

One problem of the ledged and battened door is its tendency to rack or fall away from the top furthest from the hinges. This causes binding of the leading bottom edge on the floor. To counteract this tendency braces can be installed which direct the stresses downwards towards the supported hinged edge.

This door is generally used in preference due to its greater stability and freedom from distortion.

20.05

An improvement on the ledged, braced and battened door provides for the inclusion of a framing to the perimeter of the door. This permits greater rigidity and strength as well as improving the appearance.

These doors are relatively heavy and care must be taken to select stout hinges. These are usually 100 mm cast iron butts and 1½ pairs are usually provided to reduce the strain on the knuckles and screws.

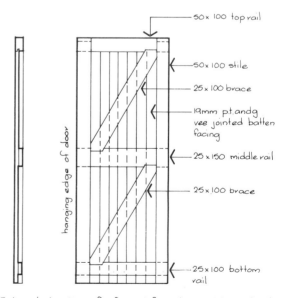

Internal elevation of softwood framed, ledged, braced and battened door

20.06

Modern doors incorporate the constructional theory of traditional doors but make use of sheet materials to replace jointed face boards. This is due not only to the need to save capital cost but also to incorporate materials which in use are more stable than timber available to the building industry today. Doors in general use fall under two main types:

(i) The *panelled door* developed from 18th and 19th century precedent which provides a wood framing, individual members secured one to another with wedged tenoned joints and with the open panels either filled in with plywood or glazed as required.

(ii) *Flush doors* constructed either

(a) with a light framing onto which a manufactured board is glued, or

(b) a solid core of timber or a fire resisting material faced with a more or less decorative plywood panel.

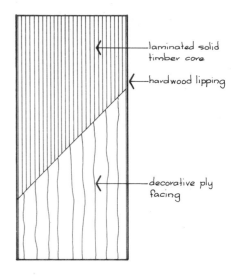

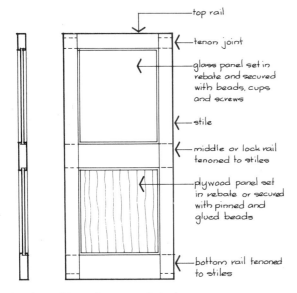

Elevation of typical two panelled door

These door types can be manufactured and obtained in a wide variety of constructions, sizes and finishes, either to the designers detail or to standard commerical production. Either door type can be provided for external or internal application.

20.07

Panelled doors are usually constructed in accordance with BS 459 Pt. 1 in a wide variety of designs and sizes. Thickness varies and generally is provided as follows:

External doors	45 mm.
Internal doors	36 mm.

The thickness of plywood used in the panel varies from 6 to 9 mm depending on the size of the panel and its location, i.e. for external or internal use. Care must be taken in manufacture to allow clearance all round the plywood with the groove to allow for possible movement and shrinkage in the framing.

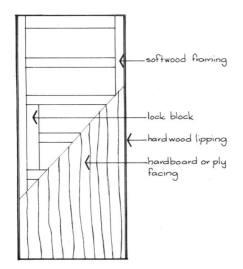

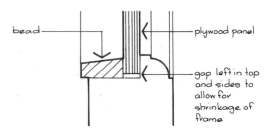

External doors are provided with standard weather mouldings at the bottom to throw water clear of the door bottom. In exposed situations these may not be wholly adequate (3.04).

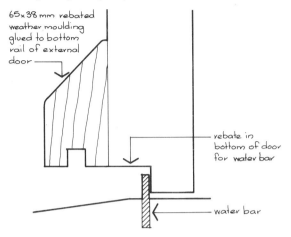

20.08

Flush doors for internal and external use are constructed in accordance with BS 459 Pt. 2, either solid or provided with glazed openings. Heights and widths conform to a limited standard range, with thickness as for panelled doors. Where glazed panels are required, the glass is fixed with beads.

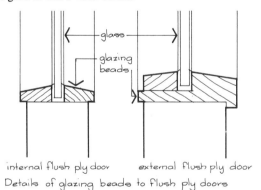

Details of glazing beads to flush ply doors

Where used externally, standard weather mouldings are provided at the bottom to throw water clear of the door bottom (3.04).

The facing for flush doors varies with the performance required. External flush doors are faced with BR grade plywood to Clause 6 of BS 1455, flush doors for internal use being faced with plywood to BS 1186 Pt. 1. In addition, internal doors are often faced with hardboard to BS 1142 which is often supplied with a factory applied primer which ensures a very good painted finish. Ply-faced doors are either supplied for painting or faced with plywood of a grade satisfactory for the application of a natural finish.

20.09 Ironmongery

The hanging of doors has been described and illustrated in section 18.04.

Ironmongery for door operation and the security of premises is very varied in design and finish but in the main falls into a fairly small number of alternatives, especially for domestic work. These may be selected from the following:

(i) Locks incorporating both a locking bolt and a spring-loaded bolt operated by means of a separate handle. These are:

(a) The mortice type which is wholly concealed within the framework of the door except for the exposed faceplate.

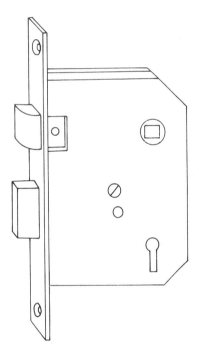

(b) The rim type which is fixed onto the inner face of the door and is wholly exposed. A good example is the cylinder night latch.

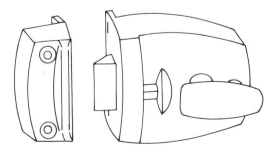

(ii) Latches, which incorporate a spring loaded bolt operated wholly by the operating handle.

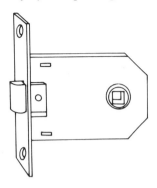

(iii) Knob and lever operated handle sets, which by means of a loose spindle inserted through the spring loaded bolt of a lock or latch enables the latter to be withdrawn from the keep fixed to the door frame.

(iv) Bolts which can either be

(a) barrel type wholly exposed on the internal surface of the door, or

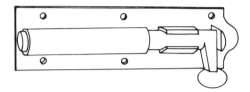

(b) flush type set into the door flush with the internal face, or

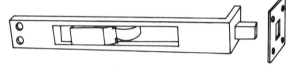

(c) mortice bolts operated by a key or handle, otherwise wholly concealed in the thickness of the door.

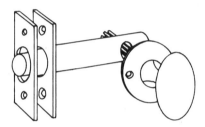

21.00 Sliding doors and gear

Sliding doors are coming increasingly into use especially where space is at a premium or door swings inconvenient. Two main types are in general use depending largely on a number of factors which include:

(i) The size of the prepared opening.

(ii) The weight of the proposed door leaf.

(iii) The threshold detail.

The two types of installation are simply

(i) 'side fixing', and

(ii) 'soffite fixing' or top hung.

The selection of the appropriate type will depend on the amount of headroom through the opening.

21.01

Sliding doors are generally of the side fixing pattern when installed to door openings. Usually a lining is provided as a plaster stop and a stout batten fixed to the side of the lintol, to which the outer track is fixed. The inner track incorporating a threaded adjustment to regulate the position of the door and the clearance between door and floor is screwed to the top edge of the door and the runners inserted in the track. A steel pelmet provides a neat cover to tract and runners.

installations are installed. The top track is screwed up to stout framing over the door opening and the runners are screwed through the back face of the door leaves, these being usually constructed from 19 mm veneered particle board.

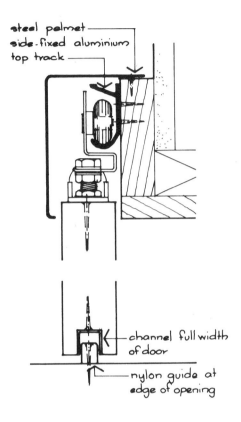

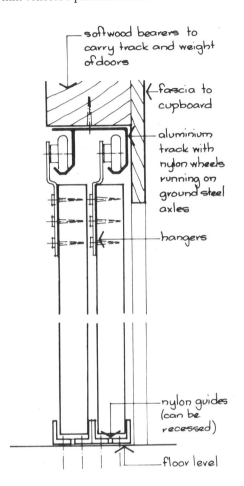

Bottom restraint to the door is provided in one of two ways, either

(i) by the provision of a continuous nylon or aluminium guide across the whole width of the opening which can cause problems as it projects above the floor finish, or

(ii) by the provision of a single nylon guide at the edge of the opening.

The construction of the track depends largely on the weight of the door leaf and is usually supplied in standard widths to suit a great variety of requirements.

21.02
Sliding doors are commonly provided to enclose fitted wardrobes and in these situations top hung

Bottom restraint is provided by nylon guides screwed either to the floor of the room or the bottom of the wardrobe which must be extended to the outer face of the framing.

21.03
Small cupboards can be fitted with sliding doors, usually 19 mm thick and carried either

(i) on top fixed aluminium track complete with nylon

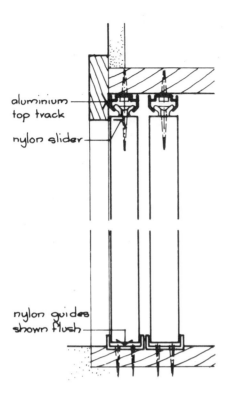

by guides in aluminium channel set flush into the top framing of the cupboard.

21.04
Glass panels forming sliding doors for cupboards can be frameless and set to run in fibre or plastic channels so long as the edges of the glass are smoothed by grinding to reduce friction and wear.

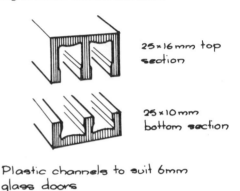

Plastic channels to suit 6mm glass doors

sliders, the bottoms being restrained by nylon guides, or

(ii) by the provision of side fixing bottom rollers running in aluminium track with the top restrained

22.00 Fire check doors and frames in timber
Certain doors are required by the Building Regulations (Part E)* to form fire resisting barriers. These doors are summarised as follows:

(i) Doors separating maisonettes or flats from common access areas*.

(ii) Doors separating houses from small private garages*.

(iii) Doors separating a habitable room or a kitchen from a stair well in small domestic structures of three or more stores (E11)*.

These doors and their frames must satisfy methods of testing and fire resistance criteria laid down in BS 476: Part 8: 1972* on the basis that if a specimen made to the specification passes the tests, all similarly constructed may be presumed to be satisfactory. In addition the door must be self-closing and while rising butts are permitted under the Regulations, few fire prevention officers will agree to their suitability or use.

22.01
In practice the Building Regulations require two types of fire resisting doors and frames, each conforming strictly to three requirements laid down in BS 476 Pt. 8. These requirements are as follows:

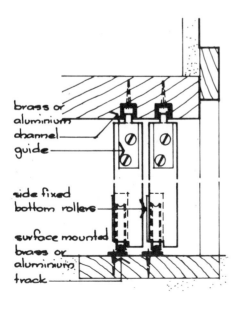

(i) Integrity. Failure is deemed to occur when cracks or other openings allow flames or hot gasses to pass to ignite a test pad of cotton wool (i.e. resistance to passage of flame).

(ii) Stability. Resistance to collapse of door during test.

(iii) Insulation.

22.02

Two types of timber door and frame are in general use providing resistance to the passage of fire. Both of these are required to conform to BS 459 Pt. 3 and provide for half-hour and one-hour periods of resistance respectively. This BS provides for standard sizes of doors as follows:

Half hour type	840 x 1982 x 45 mm. thick
	915 x 1982 x 45 mm. thick
One hour type	915 x 1982 x 55 mm. thick

The core of both types of door are similar with stiles and top and bottom rails to be not less than 38 mm thick and 95 mm wide rebated for 9.5 mm plasterboard both sides. A middle rail 168 mm wide and similarly rebated and two 45 mm wide intermediate rails complete the core construction.

The half-hour type is covered both sides with 3 mm thick plywood or hardboard, fixed and glued in a press without the use of any metal fastenings.

The one-hour type is provided on both sides with 4.5 mm thick asbestos board to BS 3536 which is then faced with ply or hardboard similar to the half hour door.

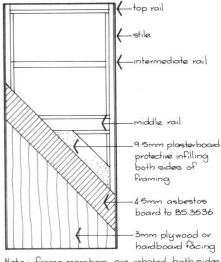

One hour type fire-check flush door

Both doors may be supplied and used with or without edge lippings. If lippings are included, they must not exceed 9.5 mm. on face with either a straight or tongued and grooved joint to the door to which they must be fixed by gluing.

22.03

Wood frames for half-hour resistance doors are required to be made to minimum requirements. The

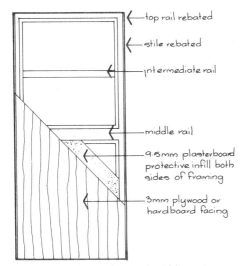

Half hour type fire-check flush door

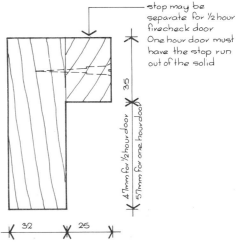

Section through frame for fire-check door constructed to BS 1567 "Wood doorframes and linings"

stop may be run in the solid or fixed with 38 mm x No. 8 countersunk screws spaced 75 mm in from the ends and then at 600 mm. centres.

There is no objection to the minimum dimensions of the frame being exceeded for practical or aesthetic reasons.

22.04

Wood frames for one-hour fire resistance doors are made to the minimum dimensions shown on the previous page, the only difference being that the stop must be worked from the solid. The timber must be impregnated with a 15–18% solution of mono-ammonium phosphate in water either before or after machining to a depth of not less than 14 mm. These fire check doors and frames may be finished in all the usual decorative finishes.

22.05

Fire check doors are hung on steel butt hinges with metal washers; nylon is not permitted. Half-hour doors must have one pair of hinges, one hour doors must be provided with one-and-a-half pairs.

While normal door handles and locks are generally suitable, for safety purposes the lock bolt should be of the straight dead lock pattern to avoid the risk of slamming in the locked position. An automatic closer must be fitted, this should be of either the overhead spring pattern or a patent pattern fixed into the rebate on the hanging edge of door and frame.

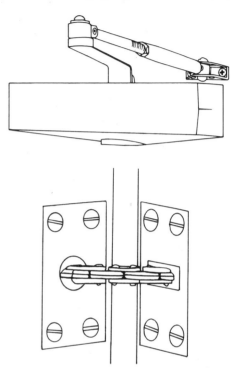

Module E Site Works, Roads and Pavings

23.00 General site works and subsoil drainage
In *Building Technology 1*, Section 9.00, the preparation of the site for building operations has been described in some detail. This preparation includes stripping the area of the works and pavings of topsoil and its disposal, either by carting away from the site or by depositing the material in spoil heaps on site for future use in the preparation of site landscaping.

23.01
During the progress of the contract, the area around the new buildings will have been used for a variety of tasks, for the positioning of site accommodation, the storage and working of materials, traffic from one part of the structure to the rest.

All these activities will have severely cut up the surface of the ground and left a good deal of broken material and waste buried just below the surface. At the completion of the actual building works all this rubbish needs to be cleared out and removed from the site before the external works are commenced.

23.02
One of the problems associated with many building sites is that of surface flooding generally caused either by the impermeability of the subsoil or a high water table. To deal with this problem on small sites it is usual to install a system of subsoil drainage whose purpose will be to:

(i) Stop surface flooding.

(ii) Improve the stability of the ground surface.

(iii) To reduce the risk of damp in foundations and especially basements.

(iv) To reduce humidity on damp sites.

(v) To improve the workability of the site soil and assist the landscaping work.

23.03
There are a number of systems in current use for the installation of subsoil drainage, each applicable for particular situations either to protect the building or to improve the site area:

(i) The *natural contour system* collecting ground water from the sides into a main subsoil drain.

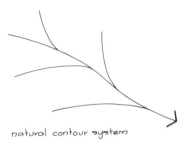

natural contour system

(ii) The *herringbone system*, either single sided or double for use on sites with little or no natural fall.

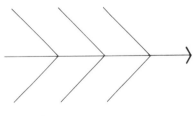

herringbone system

(iii) The *converging fan* for the drainage of specific areas falling to a natural outlet.

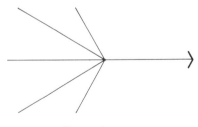

converging fan system

(iv) The *grid iron* providing subsoil drainage to relatively flat regular areas with well defined outfalls.

(v) The *moat pattern* applicable principally to buildings to protect these, especially on very wet or falling sites.

Whichever system is selected the outfall must be very

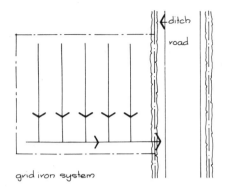

grid iron system

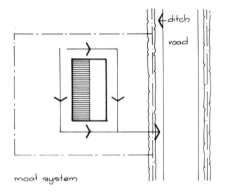

moat system

carefully considered as on its efficiency will depend the success of the whole scheme.

23.04

The procedure to be followed in the preparation of a design for subsoil drainage must include

(i) a complete and careful site survey, properly levelled with the contours plotted, and

(ii) trial holes or borings must be taken to provide accurate details of both subsoil strata and ground water level.

From this information a system may be selected best suited to the site and invert, the levels of trenches determined to provide a suitable and proper outfall to the selected ditch or natural stream.

23.05

A number of materials are in general use for subsoil drains, among which are the following:

(i) Clayware field drain pipes to BS 1196: 1971.

Manufactured to a length of 300 mm in bores of 75, 100 and 150 mm, laid end to end and butt jointed.

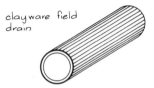

(ii) Perforated vitrified clay drain pipes to BS 65 and 540: 1970. In various lengths and bores (10.04) laid end to end with the spigot inserted loose into the socket of the pipe.

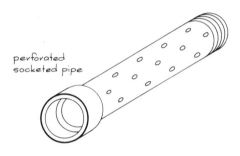

(iii) Porous cast concrete pipes to BS 1194: 1969. Manufactured in various lengths from 300–1000 mm and in bores of 75, 100 and 150 mm with rebated joints.

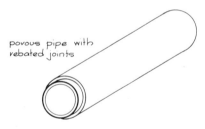

(iv) Perforated pitch fibre pipes and fittings to BS 2760: 1973. In various lengths and bores (10.06), either jointed with machined tapers or loose sockets.

23.06

Subsoil drains designed to intercept running ground water and diverting it around a building should be sited sufficiently away from the building to reduce the water table under the building to the level required. In practice this dimension will vary from 1.5 m to 5 m depending on the height of the water table, the higher the table the closer to the building will the drain need to be.

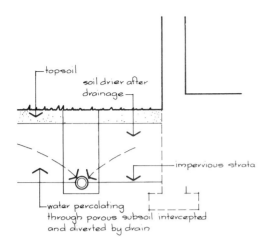

After laying in straight runs and jointed to suit the pipe material used, loose hard filling comprising large beach or hardcore broken to a 75 mm ring is filled over the top of the pipe to a depth of 225 to 300 mm and the top soil returned into the trench over the top of the pipe filling.

23.07

Having collected and diverted the flow from the building, the redistribution of the ground water must be carried out down stream from the building. Dispersal can be effected by one of three methods:

(i) Discharge into a stream or falling ditch.

The drain trench should be excavated to the approximate level of the underside of the foundation concrete, tapered in section and a width at the bottom to suit the size of pipe proposed. In most domestic work the size of the pipe will not exceed 100 mm.

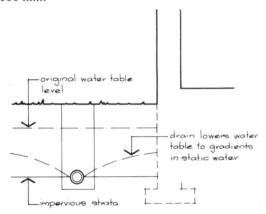

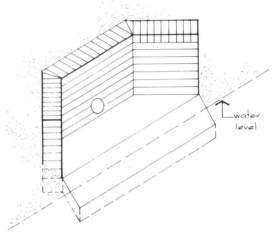

Detail of outgo or discharge point from a subsoil irrigation system to a watercourse or ditch, in brick with concrete apron

(ii) By means of a soakaway located at least 5 m away from and below the level of the building, if the subsoil conditions are suitable, e.g. rock not located close to the surface (10.21).

(iii) By means of an irrigation system equal in length and capacity to the original collection system.

23.08

With the site cleared of all rubbish and debris, levelled over and subsoil drainage installed the tops of all drain trenches should be broken in to reduce the risk of settlement, especially under pavings.

This is effected by digging over the drain runs for an area of three times the width of the trench to a depth of about 225 mm before excavating for and trimming up the bottoms of excavations for paths and pavings.

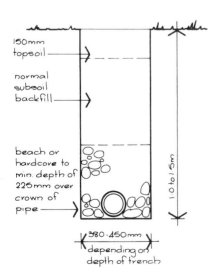

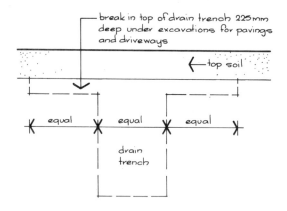

24.00 Domestic and estate footways and pavings

Buildings require some form of external pavings to provide a clean dry and solid footway. This is required for a number of reasons, as follows:

(i) To provide access from the public highway and pavements to the external doors of the property.

(ii) To provide ambulatory movement within the site boundaries to allow access from the property to areas where rubbish bins are normally stored, etc.

(iii) To provide areas for relaxation.

24.01

These paths and pavings must be permanent and require the minimum of maintenance to fulfill their proper functions. Two principal forms of construction are used, depending largely on the materials used in construction:

(i) Homogenous. e.g. *in situ* concrete and asphalt pavings

(ii) Unitary. e.g. precast concrete slabs, brick paving.

24.02

An important consideration for all paving is the removal of stormwater from the surface. Concrete and tarmacadam provide waterproof surfaces which, unless properly drained, will become water catchment areas.

With paths and small areas of pavings the surface can be set to a fall allowing water to drain off to the sides where it will be absorbed by the surrounding ground. Two methods for providing a fall are used:

(i) The cross fall method in general use for footways in which the flat area of paving is set to fall evenly from one side to another, all water being drained to the low side (usually 1 in 24).

(ii) The camber method in which the surface is arched or cambered up in the middle causing stormwater to drain evenly off on either side. This method is in general use for vehicular carriageways rather than footways where an average camber from channel to crown of 1 in 30 is required.

So long as the areas of paving are relatively narrow and the surrounding ground is well drained these methods are satisfactory.

24.03

If the area is too wide to be drained in this way or the ground soakage is unsatisfactory, it will be necessary to provide gullies to collect the water and run these to soakaways. Where the gullies occur at the perimeter of the paving simple rubble filled soakaways may be used. Where the gullies are provided

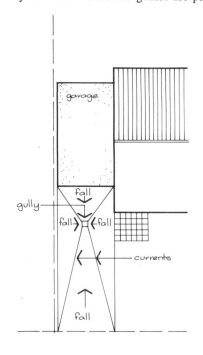

some distance from the edge brick steined soakaways need to be provided (10.21). In any event the impervious surface of the paving must be set to falls and currents to direct stormwater falling onto it into the gullies.

24.04

Where unitary surfacing is provided it is often arranged for the joints between the slabs or bricks to be opened up and filled in with sand well brushed into the joints. These then act as natural drainage channels and quickly drain stormwater into the ground.

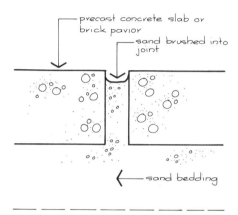

24.05 Excavations

The subsoil of a site will, to a greater or lesser degree, dictate the construction of paths or pavings. Clay subsoils react to conditions of drought, shrinking and cracking to a depth up to 750/900 mm, sometimes deeper. After subsequent rain the clay absorbs moisture and swells up. This movement can cause severe damage in continuous concrete or tarmacadam paths. Consequently where concrete is used it should be broken up into short lengths of 2–3 m where movement joints should be provided. The inclusion of a light steel mesh reinforcement helps to control and reduce cracking.

Unitary pavings, laid on granular sub base generally resist serious deformation in such subsoil conditions. On the other hand, the use of tarmacadam and similar materials over granular and well drained subsoils is perfectly satisfactory.

In any event the topsoil must be removed from the areas of paving to a depth of 150 mm and should either be set aside for landscaping or carted away from the site. The bottom of the excavation should be well rolled and consolidated to receive the sub-base material.

24.06 Sub-base materials

All pavings and footways need a well consolidated and rolled base. This is usually laid to provide the falls necessary for the removal of stormwater, the extra material used being cheaper than the finishing coat or paving. Whatever the material used, the maximum particle size should not exceed 75 mm. Materials used vary with local supply but include the following:

(i) Brick hardcore crushed to pass a 75 mm mesh and blinded with sand or similar fine material to provide a level even surface*.

(ii) Granular materials such as crushed stone, gravel or coarse sand.

(iii) Natural hoggin obtained from ballast excavations.

(iv) PFA (pulverised fuel ash) obtained from coal fired power stations.

Compaction of these materials may be either mechanical using a vibrating roller (for hardcore and granular materials) or a vibrating plate compactor (for uniformly graded materials).

24.07 Edging materials

Unitary paving materials are rarely provided with edging. Concrete requires a temporary shutter to form a fair straight edge and this is usually provided by sawn timber.

Asphalt pavings, on the other hand, require the provision of a permanent edging not only to protect the edge of the material in use but also to restrain the asphalt during laying and compaction and in use. Precast concrete edging kerbs size 150 x 50 mm are generally used, set on a small concrete bed and backed in the same material.

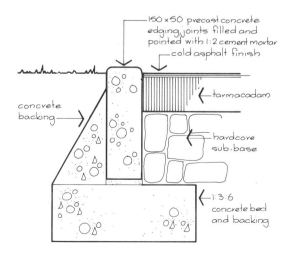

These kerbs can be obtained either in straight lengths or to a number of standard radii and are manufactured in accordance with BS 340: 1963.

24.08

In some situations the junction between a tarmacadam road and concrete paving is made using granite sets. These are very durable and are usually laid in a 1:5 dry cement mortar on a concrete bed. Cobbles are often used for decorative effects and are usually laid random in dry mortar as sets, being 'watered' in through a watering can fitted with a fine rose spray.

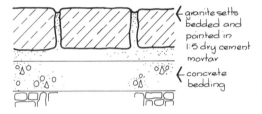

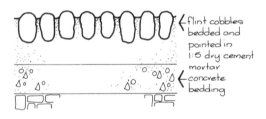

24.09 Surfacing

A number of materials are in common use for surfacing footways and paved areas:

(i) Asphalt: This paving has a fine surface texture and is obtained in either hot rolled or fine cold forms. Both are laid in one course only usually as a wearing coat over bitumen or tarmacadam and well rolled in.

(ii) Tarmacadam: Similar to bitumen macadam and much used for estate footways, consisting of an aggregate mixed with tar laid and rolled on hardcore (or similar sub-base) in one or two coats. Paths are generally either 25 or 50 mm thick and finished with a surface of fine cold asphalt of a minimum thickness of 12 mm to seal the surface. Coloured macadam in dark green or red can be obtained as a wearing course or coloured binder for paths and pavings.

(iii) Insitu concrete: Laid in bays of from 2–3 m to allow for movement and expansion and contraction and from 75–100 mm in thickness. A 1:7–19 mm cement/all-in aggregate will be satisfactory for this work. The movement joint may be left open or filled in with sand to act as a drainage channel in narrow paths or small paved areas, otherwise an expansion filler material should be incorporated in the joint.

(iv) Precast concrete slabs: Widely used especially for estate footways and for areas of domestic paving, patios, etc. Various sizes of slab can be obtained from 300 x 300 mm to 600 x 900 mm and in thickness of 38 and 50 mm. The slabs are laid in a number of ways depending on usage on a bed of well compacted hardcore or similar sub-base:

(a) On a solid bed, min. 25 mm thick of 1:4 cement and sand mortar laid semi-dry. This will provide a good base which will allow little movement and carry heavy wear.
(b) By bedding the slabs on mortar dabs on a bed of sand between 50 and 75 mm thick, well rolled and consolidated.
(c) By laying the slabs directly onto a bed of fine sand about 25 mm thick.

Joints between slabs are generally between 6 and 12 mm wide, pointed up in 1:4 cement and sand except for the last method (c) where the joints are laid open and dry or filled with dry sand.

(v) Brick paving: Usually employing special paviors manufactured for the purpose and obtained in either plain, diamond chequer or panelled varieties. These paviors can be laid either direct onto a sand bed or, if the loading is heavier onto a 1:4 cement, and sand bed.

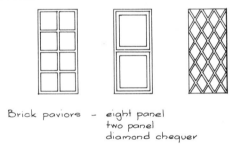

Brick paviors – eight panel
two panel
diamond chequer

25.00 Estate roads and drainage

Estate roads for housing development are always constructed to the standards set by the local authority and to the approval of the authority's surveyor. The reason for this is that the roads will, when completed and 'adopted' by the authority, become their liability for maintenance. The road surveyor should, therefore, always be consulted at the design state to ensure both the design and the specification are in accordance with the authority's standards.

25.01

Estate roads are constructed either with

(i) a flexible, granular base, or

(ii) a rigid monolithic concrete base,

depending on the nature of the subsoil and the likely use, especially with regard to commercial vehicles.

The evaluation of the bearing value and condition of the subsoil (sub-grade) is a matter for precise laboratory test from which the thickness and quality of the base materials can be accurately determined. In general, however, most estate roads use a flexible, granular base, concrete being reserved for sites where difficult subsoil conditions exist.

25.02

The widths, radius of curves and junction details must always be determined in accordance with the DOE Regulations in force at the time of the work. With regard to gradients, these should be not less than 1 in 150 and preferably 1 in 120 to ensure good drainage.

25.03

The excavation for estate roads and the evaluation of the subsoil follows closely that for estate footways (22.05).

25.04

It is necessary to provide footway crossings to give access to private garages from the public carriageway. These are of light duty construction and in most cases where the footway is less than 3 m wide and constructed of precast concrete paving flags, it is sufficient for these to be laid on a minimum of 100 mm wet concrete.

The carriageway kerb is either carried into the crossing by the provision of kerb stones to 1 m radius or 450 mm quadrants inserted. The ramped space between can then be filled in either with panelled blue bricks bedded on 100 mm concrete or the concrete brought up to level and finished with a steel float.

25.05

Where estate roads are set across sloping sites and there is a possibility of ground water being trapped in the carriageway foundations, it is necessary to provide subsoil drainage to stabilise the subsoil.

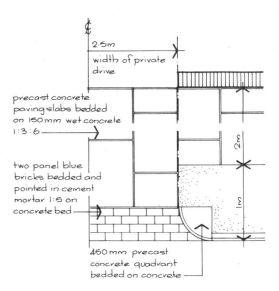

Typical pavement crossing to tarmac road with blue brick crossing

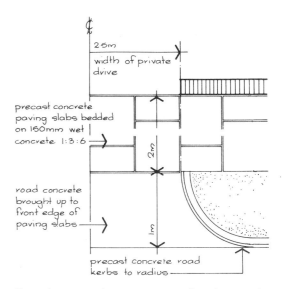

Typical pavement crossing to reinforced concrete road with concrete access

Clayware field drains either 100 or 150 mm diameter are used either laid in cross or herringbone pattern at a level below the sub-base, collected and connected to a subsoil drainage ditch laid at the foot of the slope and clear of the carriageway.

The pipe laid in this ditch is usually of porous concrete to BS 1194: 1969. This pipe must be connected to a brick inspection gulley or catchpit before connection to the sewer or soakaway to allow for removal of large particles in suspension (see also 23.05 *et seq*).

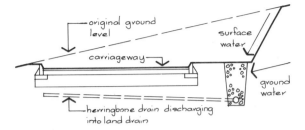

Subsoil drainage to carriageways on sloping sites

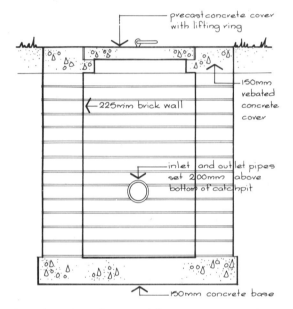

Typical catchpit for subsoil drainage constructed as for manholes (10·12)

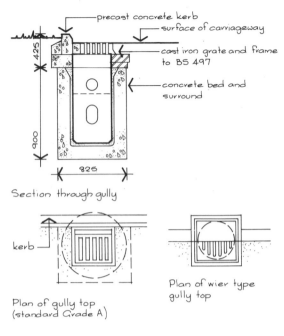

Section through gully

Plan of gully top (standard Grade A)

Plan of wier type gully top

Connection from the outlet of the gulley is to either

(i) the local authority stormwater sewer in the carriageway, or

(ii) the local authority combined sewer, or

(iii) by discharge into an existing ditch or land drain system provided in the verge at the side of the road.

25.06

Drainage must be provided to clear the carriageway of stormwater. This is carried out by installing 450 mm x 900 mm deep road gulley pots incorporating a trap and outlet in positions adjacent to the carriageway kerb where low points are provided either by design or the level of the ground contours. These gullies are either stoneware to BS 539: 1971, or precast concrete to BS 556: 1972, and are set on 150 mm thick quality 'B' concrete which is worked round the sides of the gulley to encase it completely.

25.07

The open top of the gulley pot is protected by a cast iron grating and frame, usually to BS 497: 1967, set on two courses of Class B engineering bricks in cement mortar, 225 mm wide. An alternative gulley is the weir pattern which is set within the line of the carriageway kerb.

25.08

Kerbs are provided to estate carriageways for a number of reasons, of which the following are of the greatest importance:

(i) To contain and direct stormwater to gullies provided for its proper removal.

(ii) To contain and restrain the edges of road surfaces constructed of flexible materials.

(iii) To discourage vehicles from mounting onto the footways and grass verges.

Kerbs are generally made from hydraulically pressed concrete to BS 340: 1963, in lengths of 600 and 900 mm. Three principal profiles of kerb are produced for edging, selecting being a matter for the local authority.

Curved work between 12 and 18 m is carried out using 600 mm lengths and radius work below 12 m must incorporate specially cast units.

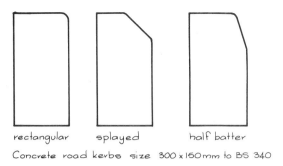

Concrete road kerbs size 300 x 150mm to BS 340

Detail of bedding to precast concrete road kerbs

The kerbs are set on a bed of concrete quality 'B' not less than 325 mm wide and 100 mm thick, the kerb being bedded in 1:4 cement and sand and the joints flush pointed in 2:1. The back of the kerb is provided with backing or haunching concrete to give additional strength to within 75 mm of the top of the kerb.

25.09

The carriageway sub-base is generally constructed of material previously described for footways (22.06). This must be rolled and properly consolidated to a finished thickness not less than 225 mm on a sub-base of fine material consolidated to a minimum thickness of 75 mm. In some areas the thickness of the sub-base can be reduced, with prior sanction from the local authority to 150 mm in thickness.

With flexible carriageways, the surface is formed from a 50 mm consolidated thickness of 38 mm nominal size bitumen macadam to BS 1621:1961, with a wearing and sealing coat of fine cold asphalt rolled to a thickness of 20 mm.

Rigid monolithic concrete carriageways are very rarely used in estate roadwork as the loads and frequency of traffic are rarely sufficient to justify their extra cost.

26.00 Reinstatement and landscaping

The completion of the building, road and drainage works is not the final stage in the provision of new accommodation, whether housing, commercial or industrial.

The areas surrounding the building have to be cleared of all waste building material and rubbish and, at the very least, the open areas of ground need to be grassed down. This work forms the basis of reinstatement and landscaping.

26.01

Before any attempt at landscaping, it is necessary to clear all builder's waste from the site, break up and clear away all brickbats, broken concrete and mortar and to thoroughly break up the hard panned areas which have been subjected to the weights of vehicles. This is best carried out by a JCB or similar vehicle using a cultivating tine to break up the surface.

At the same time all weeds and unwanted vegetable growth, bushes, and small trees should be cleared away or burnt on site. The whole area should then be lightly turned over by means of a plough and rotavator to form a fine tilth sub-base for the top soil.

26.02

If the top soil from the area of the works has been retained in spoil heaps on the site, this can now be dug out and spread over the areas to be grassed down in a thickness not less than 150 mm or to suit the new contour levels. Otherwise, top soil will have to be imported to site and dumped ready for spreading. If weather conditions are right, light consolidation can be carried out, otherwise time must be allowed for the spread soil to consolidate and compact under its own weight.

26.03

It is very difficult to cut grass right up to the edge of a building by mechanical mower. Consequently hard pavings are often provided next to the wall to

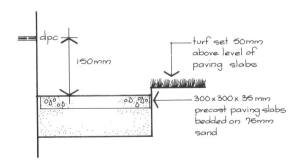

reduce hand maintenance costs and the cost of grass cutting.

A line of precast concrete paving slabs is often used, or a shallow trench filled with gravel which also helps to drain the soil next to the external wall of the building.

Flower beds with brick or tanalised timber edging can also be used to advantage. This detail will also help to soften the edges of the building.

Whichever detail is used, it must be remembered that the turf should be finished about 50 mm above all surrounding hard surfaces to enable the grass to be power cut to a firm edge.

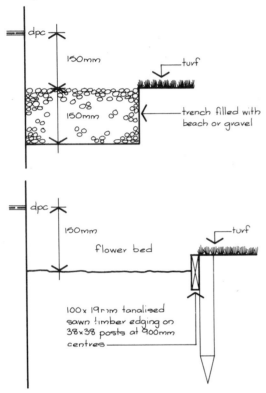

26.04

Grass turves are cut to a standard size of 300 x 900 mm and are about 38 mm thick. The surface of the ground should be lightly dressed with bonemeal at the rate of 3 to 4 oz per square metre before laying the turves.

Turves should be laid in rows with broken cross joints directly onto the top soil base and lightly beaten with a timber beater to settle them in. The joints should then be lightly brushed and filled with a fine sandy soil, with water applied through a fine sprinkler system in hot dry weather.

If any settlement occurs the turf must be taken up, the depression under filled in and levelled up with fine soil and the turf carefully replaced. Turf should be cut when required with the mower blades set high to avoid close cutting.

When turves are laid on slopes or banks they should be set diagonally across the slope and secured with 200 x 25 mm softwood pegs driven flush through the turf into the ground.

Appendix: Building Standards (Scotland) Regulations 1971-1975

This Appendix lists the references to the Scottish Building Regulations which are indicated in the text by an asterisk.

The author and publishers gratefully acknowledge the help of Clifford Large, BA, MIOB, of the Glasgow College of Building and Printing in compiling this Appendix.

Explanatory Memoranda to Scottish Building Regulations are HMSO publications designed to serve as a guide to the Regulations. They must not be regarded as an authoritative interpretation of the Regulations.

Introduction C2 Foundation and structure above foundation

Module A
Section 2.06 The Scottish Building Regulations, Part K Ventilation, lay down the following minimum ventilation requirements for houses.

Regulation	Room	Min. area of ventilator
K4	Kitchens	Opening area equal to one-twentieth of floor area
K5	Other habitable rooms	Opening area equal to one-twentieth of floor area *and* a permanent ventilation opening of 6500 mm²
K6	Bathrooms and WC's	Opening area equal to one-half of floor area with a minimum of 0.1 m². The top of the ventilation opening to be not less than 2 m above the floor.

In addition, Part Q Housing Standards requires the following: Q9, larders to be provided with a permanent ventilator with a minimum area of 3250 m² fitted with a fly-proof cover.

Section 2.07 Scottish Building Regulations do not relate minimum window size to floor area. Part L Daylighting and Space about houses, Regulation L4 requires the window to be large enough to provide a specified minimum daylight factor over a specified area of room. The minimum daylight factor required depends on the use of the room: 2% for kitchens, 1% for living rooms and 0.5% for other departments. BRE Digests 41 and 42 *Estimating daylight in buildings*, explain one method of estimating the daylight factor. Regulations L4(2) and Schedule 7, Parts I and II, permits minimum window sizes in houses to be calculated by reference to sizes given in Schedule 9, Tables 15 and 16. Regulations Q6, Heights of rooms, requires living rooms to be not less than 2.3 m in height over nine-tenths of the floor area and no part to be less than 2.1 m in height.

Section 2.09 Due to the generally higher rainfall in many parts of Scotland, door and window frames are rarely placed in the outer leaf of an external cavity wall in domestic buildings. A typical window detail is shown below.

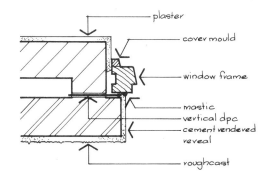

Section 3.00 Door frames rebated out of solid timber are not in general use in Scotland. Common practice is to use a planted check on both external frames and internal linings. The opening in the brickwork would normally be formed first and the frame fixed to plugs (dooks) in joints of the brickwork.

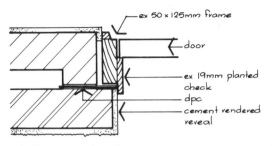

Detail of external door frame

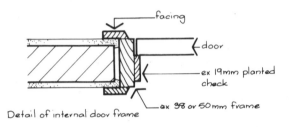

Detail of internal door frame

Section 4.01 & Section 4.02 — Part S Construction of Stairways, Landings and Balconies has similar requirements for domestic stairs. Variations occur with the following items:
(i) Minimum overlap of open treads over the tread below is 16 mm.
(ii) Domestic stairs are classed as private stairs and requirements are given in S4, column 4 of the table.
(iii) Minimum width of stair 890 mm or 600 mm if only serving one room other than a living room or kitchen.
(iv) Number of risers per flight, 3 (min) to 16 (max).
(v) No minimum riser height is given.
(vi) Minimum headroom, 2.050 m measured vertically above pitch line.

Section 4.06 — Strings are termed stringers in Scotland.

Module B

Section 7.00 — In Scotland, requirements for water installations in buildings are controlled by water byelaws made by the local Water Board under Section 60 of the Water (Scotland) Act 1946. These are intended to prevent waste, undue consumption, misuse or contamination of water supplied. The byelaws lay down requirements for materials, fittings and appliances together with minimum storage capacities for domestic buildings and the layout of pipework.

Section 7.12 — Overflow or warning pipes to storage cisterns require to have an internal diameter not less than twice the internal diameter of the inlet pipe with a minimum of 32 mm.

Section 9.02 — Part F
The same divisions are in use for appliances, but the terms Class 1 and Class 2 are not used. The equivalents are specified in F1(1) and F1(2) respectively.

Section 9.03 — No regulation deals with general structural requirements but most of the points are included in other regulations.
(i) No equivalent regulation but F3 and F21 require the chimney or flue pipe to be of suitable non-combustible material and to be properly jointed.
(ii) F5 lays down specific requirements for solid fuel and oil burning appliances and F22 general requirements for gas appliances. These are based on the need to allow free flow of gases into the air and to prevent them entering openings in the building.
(iii) F10 and 11
(iv) F10
(v) A wide range of deemed-to-satisfy specifications are given in Schedule 10 F21(1). These include the materials listed in the main text.

Section 9.04 — F13 and F15 deal with the construction of fireplace openings and hearths for solid fuel and oil burning appliances.

Section 9.05
Section 9.06 — F20A
Requirements for flues passing through floors, ceilings and roofs containing combustible material are given in Regulations F4 and F21.

Section 9.09 — See under 9.06
Section 10.02 — Part M
(i) M4 (2)
(ii) M4 (3) (c) requires the minimum size to be the greater of 75 mm or the maximum diameter of any connection to the drain.

(iii)	M4 (5)
(iv)	M4 (3) (d)
(v)	M4 (2) has a similar requirement but the phrase 'self cleansing' is not used. The accepted minimum gradient for branch drains is 1 in 40.
(vi)	M10
(vii)	M4 (5) The Regulation is expressed in general terms. No specific positions for manholes are given.
(viii)	M7
(ix)	Schedule 10, M4 (2). Regulations do not preclude other methods provided they have adequate strength and water-tightness. M5 (3)
(x)	M6
Section 10.08	M4
Section 10.12	M4 (5). No specific manhole positions are given but those listed in 10.12 would be considered to satisfy the general requirements of the Regulation. Scottish Regulations permit shallow manholes not exceeding 900 mm in depth to be constructed with walls of half-brick thickness in common brick. (Schedule 10, M8(1) (b) (2)). The roof slab of the manhole may be 100 m thick concrete.
Section 10.13	The drain within a manhole may be constructed of channels or access fittings which conform to Section 3 of *BS 539 : 1971. Additional fittings for use in Scottish drainage practice.* Access fittings may be straight pipes or bends with up to two branches. The lid of the access fitting is bedded in clay. Benching within the manhole is sloped towards the opening of the access fitting. Some local authorities do not require a sealed cast iron manhole cover to be used in shallow manholes not more than 900 mm deep. A simple concrete cover slab may be considered adequate. See illustration below.
Section 10.15	Prior to the introduction of new building regulations in 1963 many Scottish local authorities had a similar requirement. The trap used was termed a disconnecting trap and it was not incorporated in a manhole. Access to the trap is through the fresh air inlet at ground level. (See 10.16 for details) To reduce the risk of blockage and to avoid the need to prevent access to the section of drain between the trap and main sewer some authorities required the outlet drain from the trap to be larger than the inlet.
Section 10.16	Where a trap is provided between a surface water drain and a foul drain and where a disconnecting trap is used (see 10.15) a fresh air inlet is provided by extending the trap to ground or paving level with a length of drain pipe. A grating is placed over the inlet to allow air to enter but also prevent entry of matter which may block the trap (see illustration).

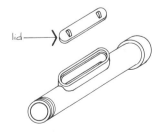

Typical access fitting

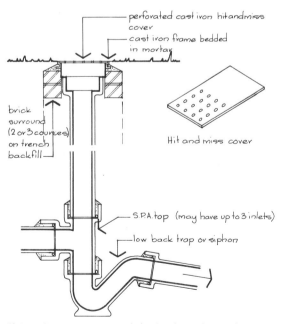

Note: the S.P.A. top and low back trap may be replaced by a Buchan trap when the drain is a straight run. Bedding for the trap is the same as for the drain itself.

Section 10.22   M4 (6) and Schedule 8 Parts I and II specify test requirements for drains. Both water and air tests are specified. In practice the commonest test used is the air test with smoke (smoke test) which has the same requirements for both surface water and foul water drains. The drain is filled with air at a pressure equivalent to a head of water of 50 mm. The test is satisfied if the head loss over a period of 5 minutes does not exceed 12 mm.
Schedule 8 Part III specifies an air test for discharge and ventilating pipes above ground. This requires a pressure equivalent to a head of water of 50 mm to be maintained for a period of 5 minutes.

Module C

Section 12.15 & Section 12.16   Many Scottish common bricks are made from shale and need to be rendered for protection and appearance. The bricks provide a moderately strong and porous background which affords some suction and mechanical key. A satisfactory bond for the tendering can normally be obtained without the need to rake out brickwork joints or use a spatterdash coat.
The traditional Scottish external rendering is roughcast or harling applied in three coats. A mix of 1:½:4 to 4½ (cement/lime/sand) or 1:3 to 4 (cement/sand with a plasticizer) may be used for the first two coats. The thickness of the first coat should be between 10 mm and 15 mm and the second coat between 8 mm and 10 mm.
Finally a dashing coat consisting of 1 part cement, ½ part lime, 1½ parts crushed stone and 3 parts sand mixed with sufficient water to obtain a wet plastic mix is thrown on the wall with a scoop. The maximum size of crushed stone may vary from 6 mm to 12 mm. Variations will be found in the mix proportions used in different parts of Scotland but the general principles are similar. On lower quality work only two coats may be used; a rendering coat and a dashing coat. Dry dash finishes are also frequently used where a more decorative appearance is desired. Mix proportions similar to those used for roughcast are suitable for the rendering which is applied in two coats. The second coat, about 12 mm in thickness, is smoothed with a straight-edge and while still soft a selected aggregate is dashed on to the surface and lightly tapped with the face of a wood float to obtain a good bond. The aggregate used may be calcined flint, spar or pea gravel well washed and drained.
Further guidance is given in Schedule 11, Part III of Scottish Regulations.

Module D

Section 19.00   The name 'architrave' is seldom used in Scotland. 'Facing' is the common term.

Section 22.00   Parts D and E
E12 (3)
D21 (2)
E9 (12) (c)
Regulation E5 may require fire-resisting doors and to be provided in other openings within houses and flats depending on travel distance to exists.

Module E

Section 24.06   Blaes from burnt out coal bings or tips is used in Scotland for a sub-base material.